THE
CHINCHAGA
FIRESTORM

THE CHINCHAGA

WHEN THE
MOON AND SUN
TURNED BLUE

FIRESTORM

CORDY TYMSTRA

 The University of Alberta Press

Published by

The University of Alberta Press
Ring House 2
Edmonton, Alberta, Canada T6G 2E1
www.uap.ualberta.ca

LIBRARY AND ARCHIVES CANADA CATALOGUING IN PUBLICATION

Tymstra, Cordy, 1955-, author
     The Chinchaga firestorm : when the moon and sun turned blue / Cordy Tymstra.
Includes bibliographical references and index.
Issued in print and electronic formats.
ISBN 978-1-77212-003-5 (pbk.).–ISBN 978-1-77212-015-8 (pdf)

     1. Forest fires–Alberta–Chinchaga River. 2. Wildfires–Alberta–Chinchaga River.
3. Fire ecology–Alberta–Chinchaga River. 4. Forest fires–Environmental aspects.
5. Wildfires–Environmental aspects. 6. Smoke–Environmental aspects.
 7. Forest fires–Prevention and control. I. Title.

SD421.34.C3T96 2015   634.9'6180971231   C2015-902847-7
                                                                                      C2015-902848-5

Index available in print and PDF editions.

First edition, first printing, 2015.
Printed and bound in Canada by Houghton Boston Printers, Saskatoon, Saskatchewan.
Copyediting and proofreading by Joanne Muzak.
Indexing by Cynthia Landeen.

The University of Alberta Press gratefully acknowledges the support received for its
publishing program from The Canada Council for the Arts. The University of Alberta Press
also gratefully acknowledges the financial support of the Government of Canada through
the Canada Book Fund (CBF) and the Government of Alberta through the Alberta Media
Fund (AMF) for its publishing activities.

To Dr. Peter Murphy, who first opened the door to this remarkable chapter of Canada's fire history, and to my wife, Hua, who lovingly supported my journey to complete this book.

CONTENTS

FOREWORD

This is a story about an incredible wildfire complex in western Canada during the summer and early autumn of 1950. Fire is frequent in the boreal forest, but this fire was truly an exceptional event. This wildfire cluster had over 100 fires that burned two million hectares—that's about half the size of Nova Scotia. The smoke from these fires was so thick, it plunged the cities and countryside of parts of eastern North America into daytime darkness. Streetlights came on, chickens returned to their roost, and people thought the end of the world was nigh. The vast amount of smoke generated by these fires pushed high into the atmosphere where strong winds spread the smoke in a band around the globe. Smoke is a serious human health issue, as recent research suggests that over 330,000 deaths every year are related to wildland fire smoke.[1] Smoke can also greatly disrupt air and ground transportation. The smoke from the 1950 fires forced Royal Canadian Air Force Flight Lieutenant Jack Jaworski to revise his flight plan from Fort Nelson, BC to Northwest Air Command in Edmonton. Two days earlier, his commanding officer flew through the fire area and encountered turbulence to a height of 14,000 feet. The cockpit filled with smoke —so much smoke that he could not see the instruments. Additionally, the smoke had rare optical effects: it caused a blue moon and a blue sun to appear in North America and Europe.

If the story ended here, this would be a fascinating book about the life of a wildland fire complex. However, this is more than a story about wildfires; it is a story of communities, neighbours, and individuals and the intersection of their lives with the life of this unique wildfire cluster. The characters in this story are real and they demonstrate the pioneer spirit of resourcefulness, self-sufficiency, and persistence. This story captures the history of western Canada during a period of settlement and change. We gain insight into the day-to-day life of people back in 1950 and efforts of individuals like Frank LaFoy. Mr. LaFoy was a forest ranger in 1950 and was responsible for fire protection for this region. We learn of his struggles fighting this fire and the bureaucracy that hamstrung him at almost every turn.

This book describes the epic battle between humans and nature with the inevitable result that we must learn to respect the power of nature, in this case illustrated through wildand fire, or prepare to suffer the consequences. Fortunately, lessons were learned and the events described in this book help shape modern-day fire management in Alberta.

MIKE FLANNIGAN
Director of the Western Partnership for Wildland Fire Science
Professor of Wildland Fire
Department of Renewable Resources
University of Alberta

ACKNOWLEDGEMENTS

There was no one location, or one person. There never is when searching for pieces of a puzzle not yet seen. I am indebted to friends and colleagues who provided pieces of the Chinchaga Firestorm puzzle. Peter Murphy's early pursuit of and reporting on the Chinchaga River Fire provided an excellent foundation to build my research from; his transcription of interviews with Frank LaFoy, Jack Grant, and Eric Huestis, and the talk by Eric Huestis to forest officers in January 1972 at the Forest Technology School in Hinton have been particularly useful for my research. While digging through material at the Glenbow Musem in Calgary, Marie-Pierre Rogeau stumbled upon Dominion land surveyor Robert Thistlethwaite's *Monument Book* and *Report and Field Journal* for his survey of the British Columbia–Alberta boundary during the summer of 1950 and winter of 1950–51. Her discovery inspired my long search for the published detailed survey maps, which I discovered accidently while wandering through the Department of Environment and Sustainable Resource Development library in Edmonton. Canadian Forest Service fire scientists Brad Hawkes and Steve Taylor, working at the Pacific Forestry Centre in Victoria, British Columbia, pointed me in new directions that yielded additional information. During a visit to the Pacific Forestry Centre, Steve showed me the early forest cover series maps for British Columbia. They too, like Thistlethwaite's survey maps, were difficult to find, and the only copies available.

Brian Stocks, a passionate, retired senior fire scientist with the Canadian Forest Service, provided information on Canada's early fire research program when "atomic bomb"-scale prescribed burns were conducted to learn about fire behaviour. During his Canadian Interagency Forest Fire Centre, National Fire Management Conversation presentation, University of Toronto professor Dave Martell challenged fire management agencies across Canada to gradually reduce fire exclusion and increase fire use. I incorporate Dave's plea for a paradigm shift in the Conclusion.

Gerald Holdsworth (research associate, Arctic Institute of North America, and retired glaciologist, Environment Canada), Robert Field (associate research scientist, NASA Goddard Institute for Space Studies), John Shewchuk (Environmental Research Services/RAOB program developer), Sharon Alden (meteorologist, Bureau of Land Management, Alaska), William Porch (atmospheric physicist, Los Alamos National Laboratory), Philip Hiscock (associate professor, Memorial University of Newfoundland), Meredith Hastings (assistant professor, Brown University), and Erling Winquest (now deceased, former aerial photo interpreter, Alberta Forest Service) all responded to my many specific questions to help complete the story.

Peter Murphy and I visited and interviewed Frank LaFoy's son Eric and his wife in Boyle, Alberta to learn more about Frank. After our visit, they excitedly contacted me to tell me that they have found the detailed colour map Frank drew of his district on the back of a large empty flour sac. Eric Huestis's daughter Donna Enger in Edmonton shared stories about her father and being part of the Alberta Forest Service family. In February 1951, Jack Grant replaced Frank LaFoy as the forest ranger based in Keg River. During several interviews, Jack recalled many colourful stories about the challenges of working as a forest ranger in Alberta's north with few resources.

Thanks are extended to the many archival staff (Alberta Provincial Archives and University of Alberta Book and Record Repository in

Edmonton, the Glenbow Museum in Calgary, the Peace River Museum Archives and Mackenzie Centre in Peace River, Alberta, the British Columbia Archives in Victoria, and Canada's National Archives in Ottawa) who helped provide access to many archival resources.

Bruce Mayer, assistant deputy minister of the Forestry Division and Emergency Response Division within the Alberta Ministry of Environment and Sustainable Resource Development, and the three anonymous manuscript reviewers provided valuable comments and suggestions to strengthen the narrative and delivery of this remarkable chapter of Canada's forest history.

Lastly, thanks to Mike Flannigan, who, despite his tireless positions as professor and director of the Western Partnership for Wildland Fire Science at the University of Alberta, kindly provided the foreword for this book.

INTRODUCTION
Land of Fire

You need big fire to create big smoke and big land to create big fire. The boreal forest in Canada has both big land and big fire. Almost half of Canada is forested; this area represents 10% of the world's forests.[1] Named after Boreas, the Greek God of the North Wind, the boreal or northern forest is the largest forest region in Canada, extending from Yukon to Newfoundland and Labrador. In 1950, the boreal forest in northeastern British Columbia and northwestern Alberta became a land of fire (Figure 0.1). The forest ranger stationed in Keg River, Alberta, mapped the active fires on September 21 as reported to him by a Royal Canadian Air Force pilot who flew over the area (Figure 0.2). This firestorm was bigger than all the others; over 100 fires burned an estimated two million hectares.

FIGURE 0.1 Location of the land of fire, 1950, northeastern British Columbia and northwestern Alberta

The arrival of fire in the boreal forest is as certain as the arrival of rain in the Amazon tropical forest. Fire is an important ecological process in Canada's northern forest. It belongs there. Some fires grow large in size, while other fires leave only small patches on the landscape. This creates a mosaic of forest patches of different age, size, shape, and composition of

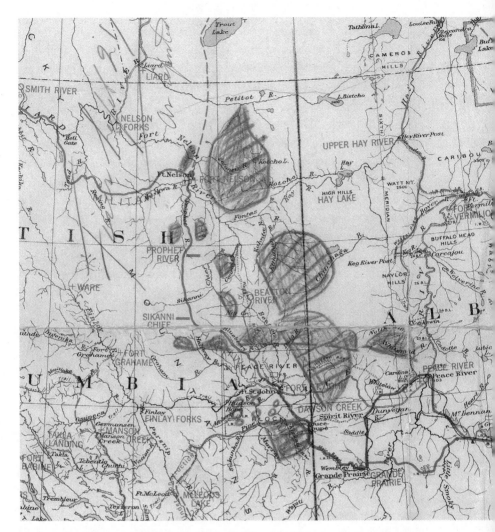

FIGURE 0.2   Active fires on September 21, 1950, mapped by Frank LaFoy as reported to him by a Royal Canadian Air Force pilot.

plant communities. From 1990 to 2010, an average of 2.2 million hectares of forest burned each year in Canada.[2] Most of the fire activity occurred in the fire-dependent boreal forest, which comprises 77% of the total forested area in Canada.[3] About 8,000 fires on average arrive annually.[4] The area burned fluctuates greatly from year to year.[5] Some years and in some places, it rains, while in other years and in other places, it burns. When it rains, the forest grows and the fuels build up. When the rains stop, the fuels dry and await the arrival of fire. This has been the natural cycle of renewal since the end of the last glacial period 12,500 years ago.

The indigenous peoples living in the boreal forest used fire. Their cultural fire regimes interweaved with the natural fire regimes. The boreal ecosystem in Canada thus evolved with both natural and cultural fire—two very different worlds of disturbance. This dance of fire changed when colonial Europeans unleashed their torch on the northern landscape. The plough may have been the primary tool in the south but, in the north, the primary tool was fire. Before the land could be ploughed, it required clearing; fire became cheap labour to ready the land.

Harnessing this powerful tool, however, was a challenge for the early settlers, who had little experience using fire. They only burned on their quarter sections of land (one quarter section is about 65 ha or 160 acres). Their desire to use fire to burn windrows and debris piles, or broadcast burn the natural vegetative cover, required drier and windier conditions than those used by the indigenous peoples. Fire was easy to start but often difficult to stop.

The early custodians of cultural fire burned different locations at different times to increase burning opportunities. This allowed for more fire across the landscape, and fewer escaped fires. The area burned, although relatively small compared to large boreal forest wildfires, provided important ecological biodiversity.

The boreal forest provided many places for fire to hide when settlers could not tame it. In Canada, the boreal forest is a huge carbon storehouse, containing an estimated 186 billion tons of carbon.[6] It was not uncommon for fire to linger and roam the boreal landscape for days, months, or an entire fire season. In peatlands, fire can sleep under the snow in deep organic layers, waiting for the arrival of the dry, spring winds. The Alberta Department of Lands and Mines was well aware of the need to check previously burned sites to ensure fire did not reappear in the spring. In their annual report for the fiscal year ending March 31, 1942, T.F. Blefgen, director of forestry, referred to controlled burns that overwintered and resurfaced in the spring as "winter fire hangovers."[7] Fire has enormous capacity to persist in the boreal landscape.

Organic matter helps to regulate the hydrologic cycle, carbon-to-nitrogen ratio, and cation exchange capacity (ability to hold nutrients such as potassium, calcium, and magnesium). It also helps to maintain soil structure and porosity and provides a site for nitrogen fixation. Approximately 30% to 50% of the organic matter in the forest floor consists of organic carbon.[8] The concentration ratio of carbon to nitrogen is important for plant growth because it affects the amount of microbial activity and, hence, the rate of decomposition.

In cool climates such as the boreal forest, natural decomposers, such as bacteria, fungi, and algae, are less efficient in the relatively cold, acidic soils. Without fire, organic matter accumulates and inhibits the growth and establishment of many plant species. Most of the nutrient budget is therefore bound in organic matter, unavailable to plants. With fire, the total site nutrient budget decreases, but the amount of available nutrients increases. This process is called nutrient cycling. Fire is the ecological key to unlocking the boreal carbon storehouse, which holds 22% of the world's total carbon reserves.[9]

Changes in soil nutrients occur during and after a fire. Nitrogen in the form of ammonium (NH_{4+}) increases. The ammonium is used by *Nitrosomonas* bacteria and converted to nitrite (NO_{2-}). The nitrite, in turn, is used by *Nitrobacter* bacteria and converted to nitrate (NO_{3-}), a water-soluble form of nitrogen that is available to plants. Nitrogen is often the growth-limiting factor on many sites.

The boreal forest is particularly unique because it is designed by nature to burn. It is truly a pyrogenic forest, where fire, as a persistent evolutionary pressure, favours plants and animals with specific survival traits. Some species are thus dependent on fire for their survival.[10] The seeds of *Ceanothus*, for example, can remain stored in the organic forest floor for several decades, waiting for fire to return. The seeds from some species may persist in seed banks up to 150 years. *Ceanothus* is a small to medium shrub commonly referred to as snowbrush. It is an important food source for elk and deer. Fire consumes the litter and upper organic (duff) layer and prepares a seed bed for the heat-resistant *Ceanothus* seeds. The seeds are scarified or heat-stimulated to germinate after remaining dormant in the soil. *Ceanothus* seeds can survive temperatures up to 93 °C for approximately 40 minutes.[11]

Ceanothus is also a sprouter. During a fire, most of the shrub is consumed. Some of the roots, however, remain alive and quickly develop suckers. The root suckers grow upward and become new *Ceanothus* shrubs. The roots survive the fire because lethal temperatures seldom occur in the lower duff layers. During high-intensity and high-severity fires, it is possible for all of the organic matter on the forest floor to be consumed, leaving only exposed mineral soil, but rarely does fire consume the entire duff layer in the boreal forest. Approximately 10% of the energy released during a fire radiates downward into the soil profile.[12]

Some pyrophilic or fire-loving species are exquisitely engineered to co-evolve with fire. *Melanophila acuminata*, a centimetre-long beetle, uses a pair of photo-mechanic infrared sensory pit organs to seek out fire. Located on each side of the middle region of the beetle's body, each pit organ houses approximately 70 infrared sensors called sensilla.[13] Each sensillum absorbs wavelengths of infrared radiation emitted from forest fires. For *Melanophila*, a distinct wavelength signature, when detected, screams "fire!" The tiny sophisticated infrared sensors can detect fire up to 80 km away. Finding this fire means finding beetle utopia. There are no enemies because they either fled or perished in the fire. The odds of this tiny beetle successfully finding a mate without fire are not good. When the beetles arrive to a fire, a mating frenzy ensues.

The female lays her eggs in crevices in the warm, charred bark of the fire-killed trees. It is a safe haven for the hatched larvae as they tunnel inward through the bark and feed on the soft, thin layer of tissue called the cambium. Since the killed or damaged trees offer no resistance to the attack, the larvae flourish.

Melanophila means "lover of blackness." Its relationship with fire is a remarkable story of how small organisms sense their environment with sensitivities that far exceed what our technologies can achieve. Each mechanoreceptor consists of a tiny cuticular sphere made of the same hard outer material that protects the beetle's body. The cuticula absorbs infrared radiation from fire at wavelengths of three to five micrometres.[14] As the sphere heats and expands, the pressure increases thereby stimulating an impulse in a single nerve cell.

Black spruce, white spruce, trembling aspen, white birch, jack pine, and tamarack are the main tree species characteristic of the boreal forest. They have different traits to survive over time in a fire-prone landscape. Jack pine trees, for example, have serotinous cones. These cones are sealed closed with resin. The heat from fire melts the resin, which opens the cones and releases the seeds. Many of the seeds survive the short exposure to high temperatures and fall to the ground. The exposed mineral soil provides a sunlit, warm, and mineral-rich site for germination and seedling establishment. The natural renewal of jack pine stands, therefore, requires fire.

Black spruce is another tree species with characteristics suited not to avoiding or withstanding fire but to embracing it. The branches on black spruce trees extend down to the forest floor thereby providing vertical fuel continuity (ladder fuels), which increases the likelihood of a surface fire climbing upwards into the tree canopy (torching) and then spreading through it (crowning). The high crown fuel load (kg/m²) contributes to sustain fire crowning. The semi-serotinous cones are located in clumps at the top of the trees to increase the survivability of some seeds after an intense fire. Black spruce can also propagate vegetatively by layering; the lower branches bend and touch the forest floor, establish roots, and grow upward to form a new cloned tree.

Trembling aspen can quickly colonize new burned areas by seeding. A tuft of silky hairs allows the tiny, light seed to be transported considerable distances by the wind. Under natural conditions, these seeds loose viability within eight weeks.[15] If a seed lands on bare mineral soil and adequate moisture is available, it will germinate. To survive, the seedling requires relatively cool and moist conditions. The high seedling mortality is offset by a very high seed production. A single, mature tree can produce up to 1.6 million seeds in a year.[16]

The Beaver, Slavey, and Sekani Aboriginal peoples, who occupied northern Alberta until the early eighteenth century, understood the ecological role of fire. They skillfully and systematically burned the boreal forest to obtain desired goods and services from the burned areas.[17] The Cree, who later moved into the area, continued the practice of burning. By managing wildlife habitat using fire, these indigenous peoples managed wildlife populations. They acquired intimate knowledge about the relationship between post-fire succession and the food requirements of the animals living in the boreal forest. When the berry shrubs flourished after burning a meadow, they knew the bears would arrive. They knew muskrat fed on the new shoots after burning the vegetation along sloughs. They also knew snowshoe hare, deer, and moose thrived on the regrowth of willow after a burn, and the increase in mice and vole populations would lure coyote, wolf, and fox looking for a small but quick meal.

The indigenous peoples in northern Alberta understood fire behaviour and risk. Fuel moisture was assessed by feeling the litter and duff and by breaking twigs.[18] Meadows, sloughs, and the grass and shrubs along streams and lakes were burned from mid-April to late May when the surrounding forest remained snow-covered or damp from the snow melt. Burning in spring minimized deep burning and damage to the plants. The spring moisture and lingering frost in the soil protected the growing tissue of grasses and herbaceous plants that lied just below the soil surface. This ensured the return of a vibrant plant community after the burn.

The early practitioners of fire use recognized the need to control it. Henry Lewis, an anthropologist at the University of Alberta, conducted many field interviews with the Native elders in northern Alberta. During one interview in 1975, a 76-year-old Cree told Lewis, "Fire had to be controlled. You couldn't just start a fire anywhere, anytime. Fire can do a lot of harm or a lot of good. You have to know how to control it...It has been a long time since my father and my uncles used to burn each spring. But we were told to stop. The Mounties arrested some people...The country has changed from what it used to be—brush and trees where there used to be lots of meadows and not so many animals as before."[19]

Burning in the spring minimized the impact when on rare occasions fires did escape. Fall burns were sometimes conducted when spring conditions did not allow for burning. These fall burns usually occurred after an early snowfall to mitigate the spread of fire into the adjacent forest. The indigenous peoples kept fire out of the mature forest because it provided the food they hunted and gathered. They depended on the forest for their survival.

With the exception of the most northern region of Alberta, the traditional hunter-gatherer lifestyle and use of cultural fire in the boreal forest were on the decline by the end of the First World War in 1918.[20] The systematic application of cultural fire soon disappeared when trapping and farming replaced hunting and gathering, and the federal government started to implement policies to eliminate fire from the landscape. When waves of settlers arrived in northern Alberta seeking free land, the indigenous peoples were told to extinguish their torches.

Despite the decline in cultural burning and perception that all fires were bad, settlers continued to use fire to burn windrows and clear the land of debris. Clearing land with only an axe and a scythe was hard work. Not all of the early settlers could afford to hire a bulldozer operator to clear their land. A 10-dollar registration fee got them a quarter section of land, but 12 ha (30 acres) had to be cleared before the government inspector arrived after three years.[21] New settlers were required to build a habitable dwelling and live on their homestead for at least six months during the three-year period. If all the conditions were met, they could apply for title to the land. Fire thus became a welcomed farmhand to help them pass the inspection.

Many young men returned to Canada seeking work when the Second World War ended. Some, such as Frank LaFoy, found opportunity in northern Alberta where the Cold War and fear of communism fed a new attention to the North. The discovery of oil also helped to increase access to this remote northern area. The few settlers who lived in the North endured isolation and many hardships. Living in the North meant surviving the North—long, cold winters and short summers fraught with black flies and mosquitoes seeking a blood meal. In 1950, the land the settlers struggled to prepare for planting became a land of fire. This was not a hardship. It was a war against a powerful force started by man but exacted by nature.

The Chinchaga Firestorm: When the Moon and Sun Turned Blue takes you on a journey through the events that happened when, in 1950, big land, big fire, and big smoke collided. It is the largest forest fire event documented in Canada, and one of the largest in the world. The significance of this event, however, is more than just its size. The 1950 forest fires generated the world's largest smoke layer. Perfect conditions allowed this blanket of smoke to be carried half way around the northern hemisphere. No other forest fire also caused the moon and sun to appear blue in colour, as it did on September 26, 1950. Those who witnessed this strange phenomenon would not forget it.

The Chinchaga Firestorm changed the boreal landscape, and man, not nature, was the sculpture in residence. This firestorm comprised an estimated 100 wildfires. These wildfires and those from the previous year suggest that humans played a major role in changing the environment. Local events, too, can have far away impacts. Although no one died, the wildfires did affect the people living in the area. One wildfire, in particular, known as the Chinchaga River Fire had lasting impacts.

Chapter 1, titled "The Lost BC Fire," introduces some of these individuals, including Frank LaFoy, the district forest ranger based in Keg River, a small settlement located 190 km north of Peace River. His duties included the protection of this community from fire. LaFoy wanted to send a crew to fight a large fire heading into Alberta from British Columbia, but Forestry Director Eric Huestis did not provide authorization to dispatch a crew, until LaFoy reported a fire threatening the village of Keg River. LaFoy's frustration with the lack of resources and the let-burn policy in effect in 1950 still resonated strongly when he was interviewed 27 years later by John Frank, an undergraduate forestry student at the University of Alberta.[22] The excerpts from this interview were submitted as part of Frank's directed study in forest fire management for Professor Peter Murphy, who started to stitch some of the pieces of one particular fire called the Chinchaga River Fire.

Frank LaFoy's interview with John Frank provided important direct source information about the Chinchaga River Fire, but it alone did not tell the entire story. Newspaper articles, letters, journals, agency annual reports, historic weather data and reports, forest cover maps, aerial photographs, conference proceedings, simulation modelling, and other interviews helped to reconstruct the sequence of events. Written accounts from local farmer Frank Jackson and Forestry Director Eric Huestis also provided different perspectives of what happened during the month of September in 1950.

Frank Jackson moved to Keg River in 1920 to homestead. In 1953, Frank and his wife, Dr. Mary Jackson [Percy], were awarded the Master Farm Family of Alberta Award in recognition of their farming proficiency.[23] Frank was also a successful entrepreneur operating a trading post and sawmill. Frank and Mary both wrote accounts of the fire.[24] He remembered the fire and the destruction of his sawmill, logs and lumber. Dr. Jackson remembered the smoke, red suns, and moose carcasses. For more than two months she could see the smoke from her home. When the fire approached the village of Keg River, Dr. Jackson described the smoke as being very thick for days.[25] When she returned to England a few years later for a visit, family and friends told her the smoke travelled across the Atlantic Ocean to Europe and changed the colour of the moon. Dr. Jackson did not doubt this story based on the amount of smoke she saw and inhaled at their homestead.

When Frank Jackson moved to Keg River, Eric Huestis was 19 years old and a new graduate from Red Deer High School in Red Deer, Alberta. Huestis worked as a teacher for one year but had loftier dreams than teaching in a one-room school in Caster, a small settlement east of Red Deer. He used the $1,200 he earned while teaching to pay for his first year at the University of Alberta in Edmonton. Pursuing his dream to become a doctor like his grandfather, Huestis enrolled in pre-medicine during the 1921–22 school year. During his sophomore year, in 1922–23, he played on the University of Alberta soccer team. They won the inter-university soccer championship that year, defeating the University of Saskatchewan by one goal.[26]

While honing his leadership skills at university, Huestis sometimes found himself in trouble. As president of the University of Alberta sophomore class, he was brought before the student court on December 7, 1922 on charges that he and his sophomore classmates were rowdy during cleanup activities after a dance. Huestis was fined 10 dollars. His encounters with the student court, officered mainly by members of the Faculty of Law, did not end there. Huestis was later expelled, ironically, for using a fire hose in the men's residence at Alberta College South, now known as St. Stephen's College. Alberta College South was the theological college of the Methodist Church in Alberta.

The expulsion did not deter Huestis. He was a savvy and energetic young man, determined to continue his studies. For the next three years, Huestis studied forestry at the University of British Columbia. In 1923–24 and 1924–25, he played on the university soccer team, and in 1925–26, he became a member of the university basketball team. Huestis excelled at sports but struggled academically. He was, by his own admission, not a very good student. Huestis continued his studies for another two years. Despite not completing his forestry degree, he forged a very successful forestry career.

In 1950, the farmer and the forester crossed paths. Huestis was then director of forestry in Alberta when fire destroyed Frank Jackson's sawmill. Both were rugged men with quiet dogged determinations and strong work ethics. They both embraced challenge and the opportunity to find solutions, and to build and fix things. It was, however, Huestis, or "Old Stone Face" as his family and friends nicknamed him, who set in motion the fire control policies and their firm adherence, thereby, according to LaFoy, allowing the Chinchaga River Fire to storm into Frank's community, continue down the valley and burn his sawmill.

Chapter 2 revisits Black Sunday and how people around the world reacted when on September 24, 1950, day suddenly turned to night. How could so much smoke travel so far and cause so much darkness? When news eventually arrived and implicated forest fires in Canada, few people believed what they could not see or smell. Chapter 3 discusses why Black Sunday stands out among the other dark days in the past.

The CBC Radio news program *Round Up*, broadcast on September 28, 1950, featured Gerrard Faye, a reporter from the *Manchester Guardian*, speaking from London, England about the blue moon and blue sun sightings from Scotland and northern England.[27] At about 1700h on September 27, the setting sun shone bright blue. Later, as the moon rose, it too appeared indigo blue. Chapter 4, "Blue Moon, Blue Sun," explains why this strange phenomenon occurred. Given that no other past fires had caused the moon or sun to become blue, why was the 1950 firestorm in western Canada different?

The environmental and health impacts of smoke described in Chapter 5, "The Big Smoke," were not well understood in 1950. A host of satellites with specialized sensors on board now assess where the smoke comes from and where it is going. Fire managers today can forecast the trajectory of a smoke plume and estimate the concentration of the pollutants it transports. Chapter 6, "The Big Wind," follows with an account of the fire weather and fire behaviour factors that contributed to the large area burned in 1950.

Using imagery from the Earth Resources Technology Satellite (ERTS-1, launched by NASA in July 1972) and sketch maps from the British

Columbia Forest Service Fire Atlas, Ron Hammerstedt, with the Alberta Forest Service Timber Management Branch, made a composite map in 1982 with the interpreted perimeters of all known fires.[28] He estimated the Chinchaga River Fire alone was about 1.4 million hectares in size.[29] In 1985, as a graduate student at the University of Alberta, I completed an independent study under the supervision of Professor Peter Murphy. Using hourly weather data, I modelled the daily forward progression of the Chinchaga River Fire. My projection fit perfectly with the mapped fire perimeter.[30]

After completing his first year of forestry at the University of New Brunswick, Peter Murphy spent the summer of 1950 working on his mother's uncle's ranch in Bridgeport, California.[31] The fires in Alberta were a newsworthy event even in western United States. An article on the front page of the September 25, 1950 issue of the *Reno Evening Gazette* became the discussion topic at the dinner table. The Associated Press story reported, "A chill blast hit a large area of the nation again today under a heavy layer of smoke from Canadian forest fires."[32] The smoke layer was reportedly about 914 metres (3,000 feet) thick and moving east over the Atlantic Ocean. The previous day, it travelled across the Great Lakes and into northeastern United States; there were reports of smoke extending as far south as the Ohio Valley. A New York weather forecaster claimed of not being aware of any other smoke pall as thick or wide as this one. For Murphy, this became the first hint of the significance of what just happened in western Canada. He returned east to continue his studies at the University of New Brunswick with the seed of curiosity planted in his mind.

In 1954, Murphy started to work for the Alberta Forest Service in the forest inventory program based in Edmonton. Two years later, he assumed responsibility for training forest rangers in the department. When the new Forestry Training School in Hinton, Alberta opened in October 1960, Murphy became the school's first director. This role provided him with the opportunity to discuss the Chinchaga River Fire with forest rangers Frank LaFoy and Jack Grant while they attended courses at the school. The seed planted ten years previously then started to grow.

In 1973, Murphy moved to the University of Alberta in Edmonton and continued to investigate the Chinchaga River Fire as part of his research program in the Department of Forest Science. Murphy recognized the need to further investigate this undocumented piece of the Alberta Forest Service's history. He believed there were important lessons to be learned from this big fire.

Did the firefighting policy fail the Keg River settlers? LaFoy's interview with John Frank 27 years after the Chinchaga River Fire and written accounts from other locals suggest it was the Chinchaga River Fire that roared into the village. The fuel, weather, and topography, and three Alberta Forest Service fire reports completed by LaFoy, indicate otherwise. Despite the considerable amount of smoke and fire around the community, the flames on the Keg River prairie did not belong to the Chinchaga River Fire; the attack came from three other smaller but explosive fires.

Politics fuels fire management organizations. When fires arrive, fire managers make decisions based on government-directed policies. Not everyone may claim victory when the battle ends; losses sometimes occur. During severe fire seasons, fire suppression expenditures often exceed the allocated base budgets. Before all of the bills arrive, program reviews and policy changes follow in hot pursuit to find ways to improve.

The 1950 firestorm in western Canada became a catalyst for change. Chapter 7, titled "Policy Changes," highlights these changes and unravels the politics and human dimensions associated with this firestorm. Chapter 8, called "Ring of Steel," follows with a discusssion of the challenges faced by fire managers then and now. Still relevant, the lessons learned from the 1950 firestorm are explored in the Conclusion, "The Big Think." This last chapter also includes a look into the future of wildfire management and the paradigm shift from reactively fighting fire to proactively managing fire and the fire environment. As discussed in Chapter 8, this shift means accepting more managed fire on the landscape and strengthening defensive positions by making communities and other values-at-risk FireSmart.[33] It is an acknowledgement that more firefighting aircraft combined with aggressive initial attack alone cannot stop all fires all the time. The past informs us of the need to find new and better ways to manage fire; otherwise, the assault of big fires, fast fires, and intense fires will be relentless and indiscriminate. The land of fire today is much different than the land of fire LaFoy knew. Increased land-use activities and values-at-risk, and climate change impacts such as increased lightning, extreme events, and longer fire seasons are already evident. Fire is also changing the fire manager's calendar. March is the new April, April the new May, and during October fires continue to arrive.

This book is a story of stories scorched with the same brand—human-caused fires driven by land-use activities, and coincident with extreme fire weather conditions and limited resources to confront taxing fire loads. In 1950, the land of fire changed; still evident are the remnants of that sculpted landscape.

The indigenous peoples of the northern boreal forest choose safe places to live, strove to protect their villages from fire, and used controlled burns to provide various ecosystem goods and services. They embraced and learned to live with fire. The changing face of fire management today necessitates a re-thinking and re-tooling to adopt a similar approach. A reconciliation of big fire events like the 1950 firestorm in western Canada allows fire mangers to think big and apply lessons learned from the past. This book guides fire management agencies down paths once travelled.

ONE

THE LOST BC FIRE

Ghost Fires and the Fire Plough

The promise of free land lured Erasmus LaFoy from Gray, Saskatchewan to homestead in the Peace River Country in northwestern Alberta in 1920 (Figure 1.1).[1] Nine years later, 18-year-old Frank LaFoy made the same journey and obtained a quarter section of land near his uncle Erasmus's homestead. After three years, Frank LaFoy gave up the land. Despite his claim the land was not good, LaFoy was a much better trapper than he was a farmer. Furs, not crops, would pay his bills. Although trapping during the 1930s was difficult because of the low prices for fur, LaFoy saved enough money to buy a trapline from Bob Bieraugle

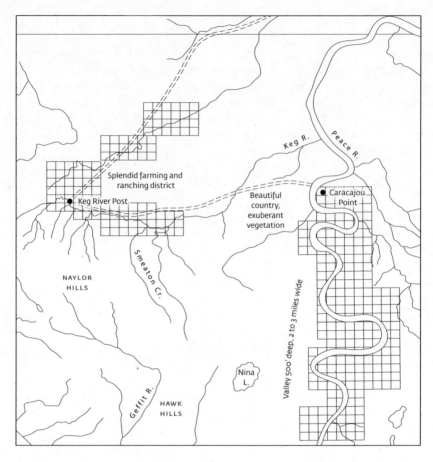

=== Wagon trail

FIGURE 1.1   Survey map of the Keg River area, 1922. In Alberta's Township System, a Township includes 36 Sections. The grids outline surveyed Sections designated for settlement. The wagon trail leads from Carcajou Point to Keg River Post and then north to Fort Vermilion. The Keg River prairie was accessible only by boat along the Peace River.

in 1935. In 1943, he sold his trapline and joined the Seventh Canadian Reconnaissance Regiment, 17th Duke of York's Royal Canadian Hussars.[2]

After serving overseas in the Second World War for three and a half years and surviving the Invasion of Normandy on June 6, 1944, Frank LaFoy returned to northwestern Alberta in the spring of 1946.[3] He was 35 years old and unemployed. LaFoy applied for a job with the Royal Canadian Mounted Police but decided not to wait six months before positions were offered. Instead, he sought employment in forestry. At that time, the field positions in northwestern Alberta were only seasonal positions; a summer fire ranger stationed for six months at Third Battle, just north of Manning, and winter game guardians stationed for six months at Little Red River, Fort Vermilion, Upper Hay River Post, and Keg River.

In 1947, the Forest Service created full-time forest ranger positions by amalgamating the winter game guardian and summer fire ranger positions. The forest ranger position interested LaFoy. He applied for the job and soon thereafter became the first permanent forest ranger stationed at Keg River (Figure 1.2), a small settlement where the Slavey and Cree First Nations once established their summer village. Keg River remains a small farming community located 547 km northwest of Edmonton (Figure 1.3).

The Alberta government paid LaFoy a starting annual salary of $1,140 and provided an additional $150 to build a house east of Keg River at the junction of the Keg River Trail and what is now the Mackenzie Highway.[4] That year, LaFoy married Anne Lorencz from Manning. Thus began LaFoy's career with the Department of Lands and Forests—a career that would take him and his family to nine different locations over the next 26 years.

LaFoy accepted the demanding duties of the forest ranger position. He enjoyed travelling in the backcountry. Like many forest rangers, he preferred the outdoors rather than the office. Avoiding and challenging the bureaucratic demands and requests from the Division Office in Peace River and the Headquarter Office in Edmonton became an unofficial skill of the pioneer forest ranger. LaFoy would again test this skill during the 1950 firestorm in his district.

In 1947, the provincial government of Alberta signed an agreement with the dominion (federal) government to establish the Eastern Rockies Forest Conservation Board (ERFCB) to co-fund and co-administer the management of the Crowsnest, Bow River, and Clearwater Forest Reserves. These reserves included the headwaters of the North and South Saskatchewan rivers. Ensuring the continued flow of precious water to the dry Prairie provinces was a priority for the dominion government. They also realized the Government of Alberta should not be burdened by shouldering all the cost of managing and protecting the forest reserves.

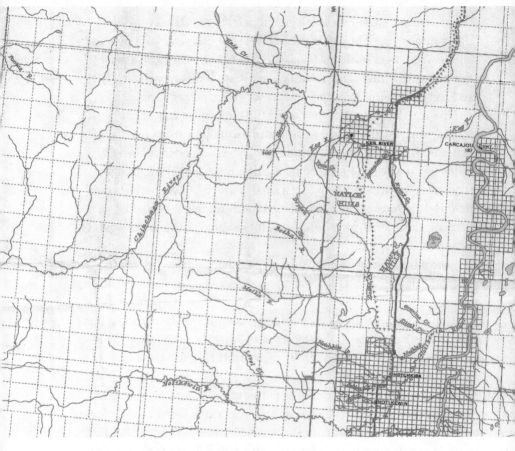

FIGURE 1.2 Map of the Keg River area, 1950. The Mackenzie Highway was completed in 1948. It is shown here as the dark line running through Hotchkiss and passing east of Keg River. The remote Chinchaga River Valley was accessible only by foot or horse. The grids outline surveyed Sections designated for settlement. Source: 1950 Alberta Motor Association map, Glenbow Museum Archives.

Several lightning fires in the 1930s underlined the perceived importance of protecting the forest reserves in southern Alberta. In 1934, a lightning fire made a major run down the Castle River Valley and burned 7,511 hectares.[5] Then, in 1936, the Galatea Fire started on the upper slopes of Mount Galatea and burned 7,290 hectares in the Kananaskis Valley, north of what is now Peter Lougheed Provincial Park.[6] Two other lightning-caused fires in 1936 crossed the Continental Divide

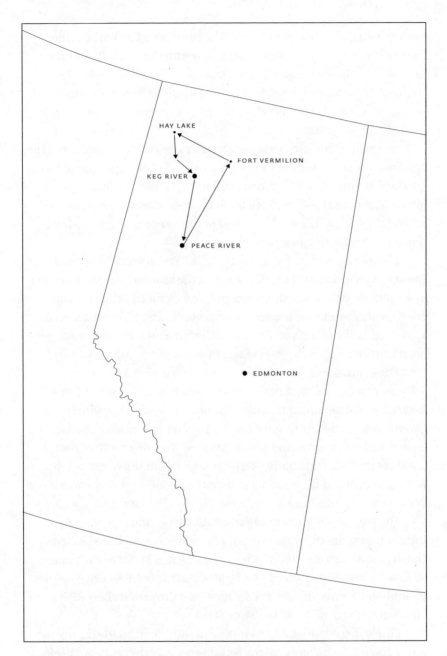

FIGURE 1.3   Regional map showing the locations of Hay Lake, Fort Vermilion, Keg River, Peace River, and Edmonton. Pilot Johnny Bourassa followed the outlined flight path to deliver mail from Peace River to Fort Vermilion, and Hay Lake.

from British Columbia into Alberta. The Pass Creek Fire burned from the British Columbia Flathead River Valley into the Castle River Valley in Alberta. This fire burned 18,568 hectares on the Alberta side.[7] The Highwood Fire burned from the Elk River Valley in British Columbia and spotted across Highwood Pass into the Highwood River Valley where it burned 24,500 hectares.[8]

Increasing land use along the eastern slopes of the Rockies, and the possibility of new wildfires causing another flood event similar to the big flood of June 17, 1897, worried the dominion and provincial governments. After the 1936 fires, both levels of government renewed discussions, which led to the establishment of the ERFCB in 1947 to address concerns about watershed protection.

The 1947 agreement formed part of the Eastern Rocky Mountain Forest Conservation Act, which mandated the board to address watershed management issues. The three-person board included a chairperson, one person appointed by the dominion government, and another person appointed by the Alberta government. The focus on watershed management continued until the 25-year agreement ended on March 31, 1973.[9]

The Cypress Hills Forest Reserve in southeastern Alberta, and the Brazeau-Athabasca Forest Reserve just north of the Clearwater Forest Reserve, remained under the sole administration of the provincial government. Collectively, all of the mountain forest reserves became known as the Eastern Slopes Forest Reserves. The area north of these forest reserves was called the Northern Alberta Forest District (NAFD). The NAFD included nine geo-administrative divisions. LaFoy was a NAFD forest ranger in the Peace River Forest Division in District no. 4.

The federal government began establishing forest reserves in Canada before the Dominion Forestry Branch was established in 1899. The Rocky Mountains Park in Alberta was created by Act of Parliament on June 23, 1887.[10] By 1911, the Rocky Mountains Forest Reserve was over 4.7 million hectares in size, and included the Crowsnest, Bow River, Clearwater, Brazeau, and Athabasca Forests.[11]

Employing qualified and well-trained staff, particularly during the war years, was a challenge for the Forest Service. Many young men left home to fight the war overseas. Those who stayed fought other battles at home. Seventeen-year-old William Harpe died on May 21, 1943 on the

Mile 283 Fire. He fell asleep on the fireguard and was run over by a bull-dozer. Wilfred Derocher and Raphael Klein, both 18 years old, died on May 4, 1944 while fighting the 15,000-hectare Moose Hill Fire in the NAFD.

Just over one year later, all timber inspectors were advised of these fatalities in a June 4, 1945 memorandum sent by Frank Neilson, the chief timber inspector for the Alberta Forest Service.[12] Neilson concluded that the three men who died were inexperienced boys who did not receive adequate safety training. He suggested the department increase the awareness of the life threatening firefighting hazards to the fire crews.

During the postwar industrial boom, low unemployment, high wages, and recruitment competition from the oil and gas sector created staff turnover problems. In the NAFD, annual meetings with forest rangers, chief rangers, and timber inspectors were held each winter. These annual two-day ranger schools proved to be an effective method for training and technology transfer. The 1949 Forest Ranger Meeting for the Eastern Slopes Forest Reserves was held in Red Deer from January 11 to 13.[13] The NAFD forest rangers met separately in Edmonton from January 25 to 27.[14] After the 1949 ranger meeting, LaFoy was reclassified from assistant forest ranger and game officer to forest officer. In February 1950, the first formal forest ranger course was held at the Banff School of Fine Arts.[15] Fifteen Forest Service staff attended this inaugural five-week course. That year, the Forest Service also issued stylish forest green uniforms. Red shoulder patches proudly identified LaFoy as a Forest Service employee.

The fire season in 1947 was a quiet one for LaFoy, with only two Class C (over four-hectare and up to 200-hectare) fires reported in the NAFD Peace River Division.[16] The following winter was long and cold, which, along with a heavy snowfall, resulted in a late spring and another relatively quiet fire season. Only 16 fires were reported in the Peace River Division during the 1948 fire season.[17] The quiet fire seasons afforded more time for forest rangers to fix and build things; LaFoy completed construction of the fire cache at Keg River in 1948.[18] Despite a light winter snowfall and dry spring, only 14 fires were reported during the 1949 fire season in the Peace River Division.[19] The 1949 fire season, however, was one of the worst fire seasons ever experienced in Alberta.

The 1950 fire season mirrored the 1949 fire season but became a pivotal year for the department. One fire in particular, the Chinchaga River

Fire, changed not only LaFoy's career but also the way fires would be fought in northern Alberta.

LaFoy called it the "Lost B.C. fire."[20] He gave it this name because let-burn fires, also referred to as non-actioned fires, were not officially assigned fire names. Fire suppression in the Prince George District in British Columbia was limited to the southern portion of the district. Any fire more than 64 km (40 miles) north of the Canadian National Railway became a "ghost fire." According to Jerry McKee, district forester at Prince George from 1945 to 1947, the Forest Service in British Columbia was not concerned about what happened in the northern regions of the province. In 1965, during the opening presentation at the 55th Pacific Logging Congress in Portland, Oregon, McKee, then deputy minister of forests for British Columbia, told the congress attendees, "For the vast hinterland we knew there were lots of fires up there but we called them ghost fires, and we didn't even bother to record them. How could you fight fire where there were no roads, no bulldozers, and nothing in the country but a few Indians?"[21] With limited access and available firefighting resources, and few immediate values-at-risk, ghost fires were allowed to roam the land-scape with no or little attempt to suppress them.

In 1977, Peter Murphy, in the Department of Forest Science at the University of Alberta, obtained a copy of the Chinchaga River Fire report (called the Whisp Fire or Fire 19) from the British Columbia Forest Service. This four-page report provides a summary of the events following reports by a nearby settler and the Blueberry Lookout Tower person, who spotted the fire and reported it to the British Columbia Forest Service on June 2, 1950.[22] Assistant Forest Ranger Jean Mitchell travelled 51 km by vehicle from the Fort St. John Ranger Station, and then three kilometres by foot to investigate the Whisp Fire, located near Whispering Pines Lake, south of Want Lake. When he arrived, the fire was an estimated 80 ha in size and burning in old pine and spruce slash on an expired timber license (no. 42255) abandoned by the Fort St. John Lumber Company.

The British Columbia Forest Service decided to let the Whisp Fire burn freely. Since the fire was heading towards LaFoy's district, he desperately wanted to assemble a crew to fight the fire as it neared the Alberta border. Despite repeated attempts, he was not able to obtain approval from Larry Gauthier, the forest superintendent, or Eric Huestis, the Forest Service director. The Metis trappers at Keg River volunteered (without pay) to accompany LaFoy to battle the Whisp Fire. They agreed to use their own packhorses if the Alberta government supplied their food. The trappers did not want to see the forest and their livelihood destroyed by the fire. LaFoy was given strict orders not to take a crew to the fire; the fire control policy in 1950 for northern Alberta stipulated no suppression action could be taken on fires located more than 16 km (10 miles) from a highway, settlement, or major river (Figure 1.2). This policy applied to Crown land in northwestern Alberta north of 52° 12' 00" latitude (north of Township 94).[23]

At 0815h on September 1, 1950, a pilot reported a fire to the Alberta Forest Service Division Headquarters in Peace River. Was this the Whisp Fire or another fire? The next day, Forest Service Ranger Robert Sears took a crew of six men to fight this fire. They travelled approximately 200 km by truck from Peace River to Clear Prairie, and then continued west for about 145 km using saddle horses and a packhorse to carry supplies.[24] Their destination was the British Columbia–Alberta boundary. At Township 88, Range 13, west of the sixth meridian, they encountered muskeg and were forced to return to Clear Prairie for horse feed. Three men, however, continued north on foot for another five kilometres. They stopped and, from an elevation of 792 m (2,600 feet), took a compass bearing of the fire. The bearing indicated the fire was still burning well inside the Province of British Columbia. The crew subsequently returned to Peace River. Since no suppression action was taken, Sears only completed the Alberta Forest Service Individual Fire Report (Form 135).[25] He did not complete the detailed report (Form 135a) because they had not reached the fire, and it was outside Alberta.

LaFoy's handwritten notes on a map that included the perimeter of the Chinchaga River Fire as interpreted from satellite imagery indicate the fire also started along the border near Want Lake, north of Fort St. John.[26] Trappers based in Keg River told LaFoy that campfires and smudges left by seismic crews working near the British Columbia–Alberta border had escaped. These fires likely grew and coalesced with the Whisp Fire.

The southwestern perimeter of the Chinchaga River Fire has two arms, which suggests the occurrence of another fire and point of origin.

This fire, mapped separately by the British Columbia Forest Service as Fire 13, started on July 29.[27] The satellite imagery suggests it spread quickly and also joined the Whisp Fire during the blowup period from September 20 to 22.

The Chinchaga River Fire haunted the 39-year-old LaFoy. He was trained to fight fire, not to let them burn freely. Working as a forest ranger was a way of life. The district became both a workplace and home. LaFoy was the lone forest ranger responsible for protecting the district. His dissatisfaction with the let-burn policy, and his ongoing challenge to change it, led to his transfer. After the fire, the department moved LaFoy to Manning.

LaFoy welcomed the change and the opportunity to leave the politics of settler-caused fires behind him.[28] Some farmers resented having a forest ranger tell them what to do and what not to do. A new fire permit system did not help to reconcile the tension between farmer and forester. It did not matter whether LaFoy was friendly or good at his job; agriculture still ruled in Alberta in 1950.[29] When oil replaced coal as the largest source of energy in Canada in 1950, agriculture's supremacy in Alberta started to erode. The oil and gas industry alone spent $150 million searching for and producing oil in Alberta in 1950.[30] New competition also came from a growing forest industry.

LaFoy had a large district to patrol, but transportation was limited. In 1949, the Forest Service had 50 light vehicles and three large trucks to service all divisions within the province.[31] In the Peace River Division, only Larry Gouche, the forest inspector in Peace River, was approved to use a government forestry vehicle. Few of the estimated 550 km of secondary roads in the province provided access to forested areas in northern Alberta.[32] Without vehicles and roads, LaFoy travelled along trails on horseback in the summer, using two packhorses to carry supplies. During the long winters, a team of three dogs replaced the horses. The horses and dogs needed fuel but they seldom broke down. When he needed to, LaFoy used his personal half-ton truck for work, but he was forbidden to submit any claims for mileage. Payment of a mileage allowance to LaFoy for the use of his own automobile for government business required authorization from Nathan E. Tanner, then minister of Lands and Forests.[33] LaFoy was not on Tanner's entitlement list for mileage allowance. In LaFoy's district, 5,000 miles was not actually that far. LaFoy could also submit receipts to claim operating expenses such as gas, oil, grease, towing, and tires.

Later that year, LaFoy received a salary increase. An Order-in-Council signed by Premier Ernest Manning on July 3, 1950 approved a salary increase for LaFoy. He then earned $180 a month.[34]

LaFoy was intimately aware of the rich forest and wildlife resources in the Chinchaga River Valley. The timber in this area was superior to the protected timber along the Mackenzie Highway. "There was trees on the Chinchaga, and what is commonly called in that country the Hay River, that two guys just no way they could reach around them," he recalled.[35] Before the fire, LaFoy tried repeatedly to convince Eric Huestis, the director of forestry, and his predecessor Teddy Blefgen, to remove the 16 km firefighting limit to protect the resources in the Chinchaga River Valley. LaFoy took great pride in knowing his district. "I knew the west country better than they [Huestis and Blefgen] knew the goddamned streets of Edmonton," said LaFoy.[36] A map hanging in the Keg River Ranger Station supported LaFoy's claim. With remarkable detail and accuracy, he drew a map of his district on the back of a large flour sac. Every trapline on the map was coloured and labelled with the name of the trapper who owned it.

LaFoy's battles to change the firefighting policy helped to convince Huestis to hire Wallace Delahey, a forestry expert from Ontario, to review both the current and the proposed forest management operations in Alberta. During his one-year assignment, Delahey flew over the Chinchaga River area to assess the timber resources. After Delahey's reconnaissance of the forest resources, LaFoy was summoned to Peace River to meet with Vic Mitchell, the forest inspector, to locate the mature timber stands on a base map. "And, that is as far as it ever got," LaFoy lamented.[37]

Before 1950, little was known about the forest resources in the NAFD. "They had so damn much timber in the south, they didn't give a damn about the north," remarked LaFoy.[38] He was right. The Forest Service focused their efforts in the south. The trees in the Chinchaga River Valley may have been big, but they were difficult to access and too far from markets. Many of the stands were also outside the 16 km firefighting limit.

In 1951, E. Fellows, the chief forester for the Eastern Rockies Conservation Area, submitted a report to the Eastern Rockies Forest Conservation Board. This report provided recommendations to strengthen the fire-detection system in the forest reserves. Fellows concluded, "It is not possible to provide protection against a conflagration except on a very broad basis."[39] He also felt that the cost of providing preparedness adequate to mitigate the impact of the eventuality of a major fire would be prohibitive. Thus, he concluded that the forest protection program should focus on prevention and the aggressive suppression of fires when they are small in size. Fellows identified four requirements. First, build and maintain an adequate system of roads and trails. Second, implement an adequate system of detection and communication. Then third, establish small, highly trained rapid response firefighting crews. Fellows felt that this specialized cadre of firefighters should be "the equal of several times their number of ordinary firefighters."[40] These crews became the precursor to today's elite "top guns" in the ongoing war of fighting fire—the smoke jumpers, heli-attack crews, and hotshot crews. Fellows's fourth requirement, and perhaps the most important to successfully win the war against fire, was the complete coordination of the first three requirements.[41]

None of Fellows's requirements were in place when the Chinchaga River Fire raced towards the village of Keg River. In 1950, the fire-detection system for the entire Peace River Division consisted of one small wooden crawl tower located at Eureka River, northwest of Hines Creek (Figure 1.4).[42] Forest rangers who climbed these ladders were not always sure they would be coming down the same way.

The lookout network expanded after the creation of the Forest Surveys Branch and the adoption of a systematic method to optimally locate lookout sites using aerial photographs for topographic interpretation. In 1954–55, a 24-metre steel lookout tower was built just west of Hawk Hills to provide views to the east and north.[43] This tower remained in use until 1976 when it was dismantled and removed from service.

The early towers were located based on the forest ranger's local knowledge of the terrain. Contemporary Alberta wildfire management staff use a computer application called VisMap to build visibility maps for each lookout tower. These maps identify what a lookout tower person

FIGURE 1.4   Forest ranger on a small, wooden crawl tower scanning the forest for fires in the Brazeau Forest. The crawl tower shown is similar to the one used by Forest Ranger Frank LaFoy. Source: Alberta Historical Forestry Photograph Collection.

can and cannot see. They also identify indirectly visible areas where fires can be detected as the smoke column starts to rise, even though the lookout tower person cannot directly see ground level. Maps that once took months to manually construct now take minutes.

During the 1920s, the Alberta Forest Service debated whether to develop a detection system based on the use of aerial patrols or lookouts. By 1928, they concluded that the construction of an extensive network of lookouts for the sole purpose of detecting and reporting fires was the most efficient and cost-effective approach. Two factors influenced this decision: the high cost associated with aerial detection, and the centric management of the eastern Rockies. Unlike Ontario and Quebec, the lack of water bodies within the eastern Rockies would necessitate the use of land-based aircraft, which require airfields. Building airfields and buying and maintaining aircraft require considerable budget expenditures. Building and maintaining lookout towers provided a more affordable solution. Subsequently, seven lookout towers were constructed south of the Red Deer River in the Eastern Rockies Forest Conservation Area (ERFCA) from 1926 to 1931.[44] Each lookout tower cost about nine hundred dollars, excluding the cost of constructing access trails and installing telephone communication.[45] Fellows's report to the Eastern Rockies Forest Conservation Board included a recommendation to increase the network of lookout towers from seven to 29.

The easier flow of forest protection resources to the eastern Rockies did not go unnoticed by the forest rangers in the NAFD. Frank LaFoy and Jack Grant, the forest ranger who, after the fire, replaced LaFoy, had fewer resources to carry out the same duties as their colleagues in the south. NAFD forest rangers also patrolled larger areas and fought more fires. In 1949, the Forest Service staffed only six forest officers north of 57°.[46] Each officer had 26,159 km² of northern forest to protect. Between 53° and 57°, fifty-seven forest officers each patrolled an average of 3,367 km².[47]

In the ERFCA, 53 forest officers each patrolled an average of 518 km².[48] Twenty-four fires were detected and suppressed in the ERFCA in 1949. Eleven of these fires were caused by lightning.[49] The chief forester's report in the *Annual Report of the Eastern Rockies Forest Conservation Board* for the period of April 1, 1949 to March 31, 1950 noted that one of the fires accounted for approximately 88% of the total area burned for the fire season. As Fellows remarked in the *Annual Report*, "Lightning was the worst enemy in 1949 and was responsible for 5,417 acres [2,192 ha] being burned. Of this, 4,760 acres [1,926 ha] burned in a single blaze, which gave real meaning to the word *wildfire*. During the early stages of this fire the rate of spread was approximately ten acres per *minute*. Such are forest-fire conditions in the mountains at their worst."[50]

Fellows's claim and bragging rights to owning real wildfires perturbed some of the NAFD forest rangers. The boreal forest is home to some of the largest and most intense fires in the world. But Fellows knew a little smoke and some burned trees in the eastern Rockies, combined with timely public relations, were good for the program. He was usually rewarded with budget enhancements after a fire season had almost gone bad.

Whether the ERFCA really had a bad fire season was not as important as who smelled the smoke. During the 1950 fire season, only six fires were reported in the ERFCA. These fires resulted in a total area burned of approximately eight hectares and a total cost of $87.58 to extinguish them.[51] Interestingly, only one Class E (> 200 ha) fire occurred in the ERFCA during the 25-year agreement. This was the 1,874 ha fire in 1949.[52]

It would take many years before the detection capability in the NAFD equalled that in the eastern Rockies. Lookout towers were not constructed in the Keg River area until 1958 (Chinchaga Lookout Tower) and 1959 (Doig and Keg Lookout Towers).

LaFoy did not like his second-class status as a forest ranger with limited firefighting resources and a policy that prevented him from fighting fires outside the 16 km limit. Each time LaFoy thought about the Chinchaga River Fire, his anger rekindled. One day, about 12 years after the fire, he decided, "I'm going to fix it for good."[53] He took his diaries from 1946 to 1952 and burned them in a barrel. This was a tragic loss. Included in the diaries were his weekly accounts of the status of the Chinchaga River Fire.

Fighting fire is akin to fighting a war—intelligence is the key to success. LaFoy was kept informed of the progress of the Chinchaga River Fire by receiving the best intelligence available at that time—aerial reconnaissance. Once a week, pilot Johnny Bourassa flew from Peace

River to Fort Vermilion and then across to Hay Lake to deliver mail. On his return flights, he detoured south to reconnaissance the fires before flying to Keg River. Bourassa flew directly over LaFoy's house and dropped a small sack containing information about the fires, in particular, the Chinchaga River Fire.

Bourassa worked for Peace River Airways, flying a twin-engine Avro Anson built by de Havilland. He sometimes flew the Hudson's Bay Company fur inspector from Peace River to Hay Lakes. He would return about four days later to pick up the inspector. During these trips, LaFoy received reports from Bourassa twice a week. One week, Bourassa made three drops. "It was getting a little bigger, little bit bigger and a little bit bigger. Then when the wind hit she just—that was it. She just covered country in no time flat," recalled LaFoy.[54] The big wind arrived at about 1000h and by 1600h the settlers were riding around the prairie to extinguish burning spruce cones with wet gunny sacks.

On Armistice Day, November 11, 1950, Bourassa flew over LaFoy's house and threw out a small sack containing his last update. His notes were brief: after the big rain, only small smoldering fires remained. The deep peat in the muskeg would, however, continue burning for several years.

On May 18, 1951, Bourassa, then working for Yellowknife Airways Ltd., transported a team of botanists from Yellowknife to Bathurst Inlet. He flew the company Norseman Mark V aircraft; chief pilot Ernie Boffa made the same trip earlier, flying the Bellanca aircraft. At Bathurst Inlet, they planned to switch aircraft; Boffa would take the Norseman and continue north with the botanists and Bourassa would return to Yellowknife flying the Bellanca. These plans changed when they tried to start the Bellanca at Bathurst Inlet. The batteries were dead. Boffa removed his maps from the Bellanca and decided to fly north in the Norseman. He not only had his and Bourassa's maps but also the rifle that Bourassa left in the Norseman. With help, Bourassa managed to hand-crank and start the Bellanca. He left alone on his return flight to Yellowknife with no maps, no rifle, a faulty compass, and no radio communication.

Bourassa got lost, ran out of gas, and was forced to land on Wholdaia Lake, about 520 km southeast of Yellowknife. The note

Bourassa left in the plane indicated he planned to walk in a north-west direction hoping to reach Reliance or Stark Lake.[55] Bourassa never arrived.

LaFoy recalled another pilot who flew regularly from Fort St. John to Fort Simpson. On one of his trips, this pilot flew from Fort St. John to Peace River to service his plane. When he encountered Johnny Bourassa at Peace River, he drew a map showing all of the fires he observed while flying from Fort St. John to Peace River. He told Johnny, "it's just like a big fire bed underneath," and "at night it is all a fire of hot coals."[56] Bourassa recorded this information, and the next time he flew over LaFoy's house, he dropped another sack tied with a handkerchief and filled with new intelligence. The hand-drawn burned areas on the map, when digitized, suggest that the total area burned was 2,963,000 ha.

Bourassa's reports heightened LaFoy's concern that his community was threatened from the fire. The first time that local settler Frank Jackson noticed smoke at Keg River was about the end of June. Coincidently, the politicians arrived at the same time. Floyd Gilliand, member of the legislative assembly (MLA) for Peace River and Solon Lowe, the Dominion member (MP) for Peace River, visited Keg River on July 26, 1950. It was Lowe's first trip north to this part of his riding. Keg River did not have a mayor at that time, but if it did, it would have been Frank Jackson. When politicians visited Keg River, their tour usually included a stop at the Jackson homestead. During this trip, Frank Jackson and his wife, Mary, hosted a lunch for their guests.

Gilliand, Lowe, and Jackson were upstairs washing their hands just before lunch when Lowe looked out the west window and noticed smoke. Smoke worries politicians. If Lowe could see and smell smoke, so could his constituents. Although the smoke came from a fire far away, he knew there may be political fires to extinguish when he returned to Edmonton.

Assuming the fire was in Alberta, Lowe asked Gilliand if it had been reported. Gilliand also did not realize the fire was still in British Columbia. He indicated the fire must have been reported because pilots flying from Edmonton to Norman Wells, NWT would no doubt have seen and reported the fire. "Do you think it will come here? It seems a long way away, 100 miles—that is a long way," Lowe asked Jackson.[57] Jackson replied, "Yes, 100 miles or more."[58] Jackson also reminded Lowe that the current firefighting policy meant nothing would be done to try to contain the fire. The homesteaders and trappers would be left on their own to battle the fire. Jackson was correct on both points.

The fire travelled northeasterly, burning the mature spruce forests along the Chinchaga River Valley. The east perimeter of the fire followed the top of the Clear Hills. The fire burned with very high intensity and severity. In many areas, the entire forest floor, including tree roots, was consumed. During the winter following the fire, oil companies moved into the area to continue exploration activities. There were reports of 18-ton D6 and 40-ton D8 bulldozers dropping into three-metre pockets of ash.[59]

While working in the Peace River Forest in the late 1950s, Harry Edgecombe observed the impact of the fire while flying over and hiking within the burned area. He particularly remembered the severity of the fire in the Fontas Tower area. Dense forested slopes covering an area of approximately 100 km² were severely burned, leaving large areas of only white mineral soil. In 1965, when the Fontas Tower was built, most of the area still had not revegetated. Edgecombe reported, "During our tower construction in the area, the fire outline was visible from aircraft for as far as the eye could see from an altitude of 5,000 feet [1,524 m]."[60] Communications technician Ron White installed the antennas at the Keg Tower when it was built in 1959. He recalled having to fly in poles because no trees were available in the area to use as masts.[61]

LaFoy returned to the Chinchaga River Valley the winter following the fire to view the aftermath. In early December 1950, while accompanying trapper Charlie Chase to help salvage the gear he left in his cabin and on his trapline at the junction of the Little Hay and Big Hay Rivers, LaFoy couldn't even find any tree roots in the ground. The fire consumed everything. The cabin somehow survived, but all of the surrounding large spruce trees vanished. LaFoy said it was hard to comprehend how such big timber could be completely consumed. "When the fire hit those big spruce,...they just exploded," Chase told LaFoy.[62]

Chase had not been back to his cabin since he rode out to Manning once the Chinchaga River Fire abated. The ride to Manning was a long and difficult journey because the fire contaminated the water in the rivers and creeks. He lost his equipment, dogs, and packhorses. Chase feared his trapper neighbour, K.H. "Slim" Hanson, died in the fire. When Chase reached Manning, he left with pilot L.E. Clarahan to fly over the burn and search for Slim. They spotted Slim at his cabin without any apparent injuries. After dropping food and supplies, Chase and Clarahan flew back to Manning.[63]

LaFoy was familiar with the desired method of attack to fight fire. The second edition of the *Forest Fire Fighters Guide*, published in 1949 by the Woodlands Section of the Canadian Pulp and Paper Association, advised fire bosses like LaFoy first to "Have a definite plan of attack": "The point of attack on any fire is the front or head thus stopping the advance of the fire. However, sometimes it is necessary to attack from the flanks or sides first as the heat at the fire front is so intense men cannot work there. With an attack on the flanks every effort is made to pinch in the flanks and narrow the front."[64]

LaFoy did not need to develop a plan of attack, however. His community was under siege from a large, high-intensity forest fire. He needed a plan of defense. When the big wind hit the Keg River prairie, and it began to rain fire on the village, he decided to fight the fire with fire, by building a bulldozer guard and starting a backfire. The 1949 *Forest Fire Fighters Guide* discussed the use of backfiring as a method of fighting fire. Fire bosses were nonetheless cautioned: "It is an extremely difficult and dangerous practice unless the fire boss has had considerable experience in fire behaviour. It is only recommended for use as a last resort to protect permanent improvements."[65] This guide also advised to attempt backfiring only from a good natural or constructed fire line.

When fire was reported about 32 km west of the village of Keg River, LaFoy hired Frank's son Louis Jackson to take his small gas bulldozer and build a fireguard (Figure 1.5). This guard was built about 1.6 km from the fire but 14.5 km outside the 16 km firefighting limit. The fireguard saved the community, but Director Huestis refused to pay Louis Jackson.

Huestis followed policy religiously and expected everyone else to do the same. The rules applied not only to the Forest Service field staff, but also to their families. Government policy stipulated that Huestis was forbidden to transport family members in his government vehicle. On one particular day, Huestis's wife, Ivy, called him at work and asked if

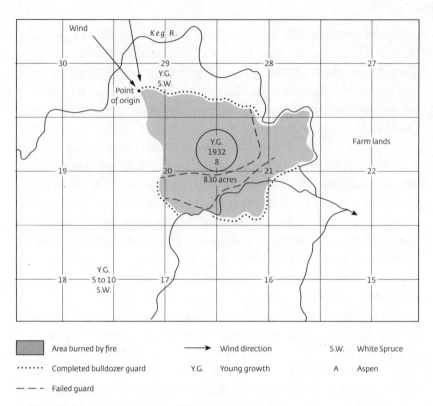

▬ Area burned by fire	⟶ Wind direction	S.W. White Spruce
⋯⋯ Completed bulldozer guard	Y.G. Young growth	A Aspen
— — - Failed guard		

FIGURE 1.5 Map of the Rat Fire completed by Forest Ranger Frank LaFoy and included in the Rat Fire Report. The numbers on the map refer to the Alberta Township System Section number. There are 36 Sections in a Township and each Section is divided into four quarters. Approximate tree heights and stand origin dates are also given.

he could pick up their youngest daughter, Donna, and bring her home for lunch. The walk from Westmount School to the Huestis home was about 1.6 km, but it rained heavily that day in Edmonton. Huestis drove to the school, handed his daughter a rain jacket, and drove back to work. On another occasion, the Muttart Lumber Company gave the Huestis family a large ham. The girls looked forward to gorging on the ham for dinner but when Huestis arrived home and saw the ham, he promptly returned it to the company. Government policy did not allow employees to accept gifts.[66]

Huestis's unwavering adherence to policy generally benefitted the Forest Service, an organization built upon a paramilitary command

and control structure. He sometimes relented, although rarely. But Louis Jackson eventually received payment for his work building bulldozer guard.

Setting a backfire was considered a legitimate strategy when no or limited firefighting resources were available. Interestingly, the historical data suggests that backfires were applied more frequently during the 1950s than during the following two decades. The increase in technological power, in particular, the use of aircraft, and increased concerns from the forest industry about potential timber losses from backfires closed the door on its use. At the start of the twenty-first century, backfires started to make a comeback, not because of limited firefighting resources, but rather because of the ineffectiveness of these resources to contain a firestorm. A firestorm is one or more very intense fires, usually large, that generate strong in-draft winds, a towering convection column, and extreme fire behaviour such as the occurrence of firewhirls or mini-tornadoes. During a firestorm, indirect attack by setting a backfire is not only the last choice; it can be the most effective strategy to manage these fires. Fire management agencies today have a much greater understanding of the fire science related to using backfires. The twenty-first-century forest industry also better understands the benefits of using indirect attack to suppress fire and reduce timber losses.

With the blade set at an angle, Louis Jackson ploughed ahead with his small bulldozer, removing the vegetation and organic layer. The Metis trappers followed behind him setting a backfire by hand to burn out the vegetation between the guard and the approaching fire.

The guard held the main fire but burning spruce cones flew across the guard and caused problems for the farmers. According to LaFoy, "That is actually what saved Keg River...was the fire guard I got, but the other thing was the damn spruce cones was a [sic] falling in the hay fields and meadows and stubble fields. You had to have somebody...well, the farmers pretty well looked after that...they had a saddle horse, and wet gunny sacks. They'd see a smudge out in the field or out in the stook and they would go out and put it out. That lasted about three days, I guess. These damn cones were falling. But the one day was the worst."[67]

LaFoy recalled that the Armstrong family was in their cabin beside the Peace River at La Crete when the big wind hit. They watched as glowing cones fell into the river.[68] After the fire, LaFoy heard about an Aboriginal elder who put out a fire 16 km north of Carcajou, which also started from burning spruce cones falling from the sky. This suggests burning embers carried aloft travelled long distances. LaFoy said the ember attack lasted three days, but one day the wind blew 65 km/h for about 12 hours.[69] At night, the embers resembled falling stars.

LaFoy drove around and advised everyone in Keg River and Paddle Prairie to protect their own property. It was harvest time and most of the grain crops were cut and stooked—that is, the bundles of cut wheat, oats, and barley were standing on end in the fields, left to dry. As the fire advanced eastward, Dan Mudry, a hired hand, frantically moved stooks in the northwest quarter section on the Jackson homestead so he could plough a fireguard east of a large bluff located in the northwest corner of the northwest quarter section (Figure 1.6).[70] As Mudry ploughed the fireguard, he watched as various animals ran ahead of the fire and crowded onto the bluff. The animals and the Jackson homestead survived the fire. Mary Jackson's daughter, Lesley Anne, ploughed another fireguard in the southwest quarter section to protect their log home and barns. The Jacksons were not worried about the fire approaching from the south since their neighbour recently ploughed his fields. The main threat would be from a fire advancing from the west side of their quarter section.

When the fire reached the Trosky homestead located in Section 32, just west of the bluff on Jackson's Section 33, it turned north into Frank Gordon's field (Section 4) located north of the Jackson homestead. The fire burned Gordon's farm equipment and two bins full of grain from the recent year's harvest.[71] The firefighters managed to save Gordon's log house, but as told by his son Joe, Frank suffered a heart attack, was hospitalized, and did not return to the homestead he built upon arriving at Keg River in 1929. Saving money while working on a homestead in northern Alberta was a challenge, particularly during the Depression and war years. The loss of Frank Gordon's harvest in the fire further strained him financially. He would not be able to bring his wife from Poland to Keg River. Dr. Mary Jackson's recollection of the fate of Frank Gordon suggests the fire also took an emotional toll on him. According to her, Frank went insane after the fire and died in Ponoka, a small town located 95

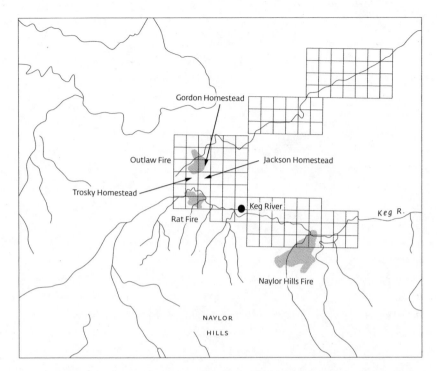

FIGURE 1.6   Location of the Rat, Outlaw, and Naylor Hills Fires (in gray) and the Gordon, Trosky, and Jackson homesteads. Grids outline surveyed Sections designated for settlement.

km south of Edmonton where the Provincial Hospital for the Insane first opened in 1911.[72]

This was not Trosky's first encounter with fire on the Keg River prairie. He lost haystacks during a fire in the 1920s. Trosky ploughed five or six strips around his stacks of hay and then another five or six strips 7.6 m (25 feet) out from these stacks.[73] To further strengthen the fireguard, he burned the grass between the two sets of strips. Despite Trosky's efforts to protect his two haystacks, they both burned. The ploughed strips did rearrange and partially bury the hay stubble, but the fire burned under the ploughed strips and crawled across the fireguard by burning patches of the hay stubble. Once the first haystack caught fire, a storm of embers ignited the neighbouring haystacks, including one belonging to Frank Jackson. Jackson estimated he lost about 27,215 kg (30 tons) of hay.[74]

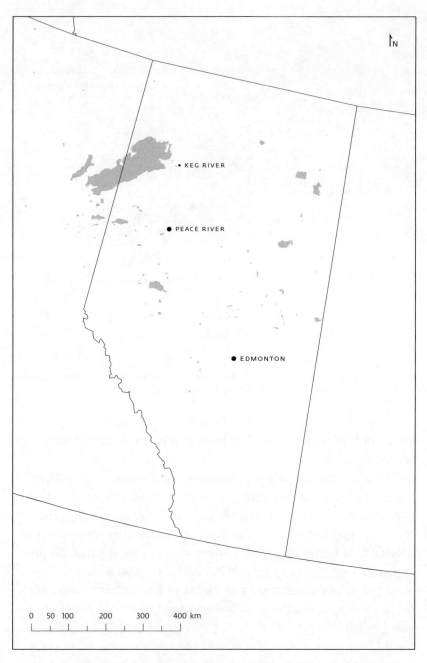

FIGURE 1.7 Mapped fires from Alberta's fire history database that occurred during September 1950 in Alberta and northeastern British Columbia.

Although none of the 1950 fires reached the Jackson homestead, many embers fell and threatened to burn their log cabin. Mary Jackson vividly remembered one of these days: "The house was full of smoke, but we had to keep the windows open to get some air, there seemed to be so little oxygen in it. Ashes covered everything, floors and furniture were all grey."[75]

The recollection of events by people who lived in the Keg River area in 1950 is not all consistent, which is understandable given that fire and smoke were everywhere and memories fade over time. The written accounts refer to the Chinchaga River Fire burning eastward down both sides of the Keg River, but how far did the fire travel? Did the fire cross the Mackenzie Highway and reach the mighty Peace River?

Five separate Forest Service fire reports completed by LaFoy in 1950 tell the story of what happened when fire overwhelmed the village of Keg River and the surrounding prairie. LaFoy burned his diaries, the report he completed on the Chinchaga River Fire, and possibly the other five fire reports, but the Alberta Forest Service fortunately kept copies of those five reports on microfilm. The fire reports confirm the Chinchaga River Fire did not reach the village of Keg River, but its embers likely did.

The fire just west of the Keg River Post was the Rat Fire (Figure 1.6). Natives camping by Duck Slough reportedly started this fire on September 11 at 1500h.[76] The name Rat refers to muskrat—1950 being a bumper year for these rodents. The Alberta Forest Service reported five active fires in the province on September 11. Two bulldozers and 10 men fought a fire burning in heavy windfall south of Sturgeon Lake, located 400 km southeast of Keg River. No resources were sent to help LaFoy protect his community from the Rat Fire.

When T.H. Bowe, a local settler, saw smoke from the Rat Fire, he went to investigate. His efforts to extinguish the fire failed because sustained strong southwest winds fed the fire. Bowe subsequently notified LaFoy by telephone and LaFoy arrived on-site on September 12 at 0400h to investigate. He couldn't take any action until he notified the Peace River Division Office because the fire was outside the 16 km firefighting limit.

On September 13, a Forest Service inspector arrived at 1630h to assess the situation. When sanctioned suppression activities began an hour later, the Rat Fire had grown to 12 ha in size. LaFoy hired fire crews consisting mainly of local trappers because they lived and worked in the bush. Most of the 48 trappers who called Keg River home were also single. The 35 cents per hour wage for fighting fire helped pay their bills. Many of the homesteaders, too, fought fire to protect their homes, barns, family, and crops, but without pay.

By 2000h, 14 hired trappers established a guard on the south side using shovels, axes, and water bags. The next day, the wind direction changed from northwest to southwest. The fire jumped the south guard at about 1030h when 35 km/h winds started to blow. Another guard, built using a single tractor and plough, held until strong winds hit Keg River on September 20. The fire again ran with the wind.

The fire crew successfully set a backfire along 695 m of fireguard but lost over one-third of the total 5,989 m of fireguard constructed to contain the fire. To push a 1.6 km fireguard through mature spruce stands, LaFoy brought in a bulldozer. The fire crew was divided and deployed to two separate locations to stop the fire from advancing into the village. They held the fire until the arrival of rain and snow on October 4.

LaFoy's last sentence in the "Remarks" section of the fire report describes the challenge he encountered: "High sw [southwest] winds caused lots of trouble with fires running away."[77] The use of the word "fires" indicate more than one fire burned out of control. LaFoy also added the words "no location" just outside the remarks box on the report form. This suggests too much fire and too many embers happened too quickly for one forest ranger to manage as he also tried to map all of the fires for the Peace River Division Office.

"100 Bush Fires Ravaging North"

On Thursday, September 14, seventeen fires were reported burning in northern Alberta. Smoke from these fires, as well as three large fires in Alaska and other fires burning in British Columbia, blanketed Edmonton. The next day, flights in and out of Blatchford Field, one of the busiest airports in the world, located north of the city centre, were delayed. The smoke continued moving south engulfing the cities of Red Deer, Calgary, and Lethbridge. According to Alberta Forest Service officials, the largest and only fire of concern is a fire burning in the Lac La Biche area in northeastern Alberta. Ten firefighters worked hard to stop this fire from burning timber resources in the adjacent mature forests.[78]

During the next five days, 10 new fires arrived. The Alberta Forest Service dispatched 16 bulldozers and 135 men to battle fires throughout the province. The public was reassured that these fires burned primarily in old burns (areas previously burned). The fire located near Sturgeon Lake was the only fire burning in commercial forest, or what the Forest Service referred to as "green timber." Despite the continued drying and increased threat of fires, the Alberta Forest Service considered the situation to be under control. The alarm bell subsequently remained silent.

On Monday, September 18, high winds blew up a fire smoldering in peat muskeg for several years near the Village of Newbrook, located about 120 km north of Edmonton.[79] The firefighting activities on the Newbrook Fire were an unorganized effort of 200 residents from Newbrook and the surrounding area. They fought desperately in choking smoke for three days and nights to protect their properties from the fire. The North American Railway parked a railway tank car full of water at Newbrook to help with the suppression efforts to prevent further property losses.[80]

Another fire that smoldered for about a week near the Village of Wanham, located 400 km northwest of Edmonton, grew 7,800 ha on Wednesday, September 20 under the influence of gusty 64 km/h winds.[81] Of the 27 fires burning out of control, the Wanham Fire caused the most trouble for settlers. When this fire destroyed three farms located about six kilometres south of Wanham, in the Birch Hill area, the province could no longer ignore the pleas for help. Farmers J. Benson, Art Tansen, and A. Simpson lost their homes, barns, equipment, crops, and animals. Everything they owned was destroyed and uninsured. Many farmers scrambled to stockpile whatever they could carry to the summer fallow fields, and then fled. Betty Wells remembered seeing their furniture standing oddly in the middle of a ploughed field. Their empty house, constantly soaked with water, survived the inferno. Her neighbour, Albert Olson, struggled to protect his homestead, but lost his barn.[82]

The province initially sent two bulldozers to assist the 35 men, most of them farmers, fighting the Wanham Fire, but these resources were no match for the fire. The firefighting efforts on the Wanham Fire were unorganized. The *Edmonton Journal* reported that "each farmer is too busy looking after his own farm and possessions to worry about his neighbor."[83] On the Sturgeon Lake Fire, 36 bulldozers provided support for the 135 men fighting the fire. The Wanham and Sturgeon Lake Fires were the most serious of all the fires burning in the province, but a new fire in the Fort McMurray area drew the attention of the Alberta Forest Service, who rushed to send bulldozers and men to this fire.

On September 20, the Rat Fire jumped a ploughed fireguard, forcing LaFoy to set a backfire to protect the village. The winds diminished that evening and, on Thursday, provided a reprieve for the tired firefighters and farmers to harvest their crops while flames encircled their fields. Delbert Wells and L. Christian dropped unconscious from exhaustion while harvesting their crops during the night in thick smoke.[84]

The occurrence of another fire called the Outlaw Fire, northwest of the village of Keg River, corroborates the historical accounts of fire on both the north and south sides of Keg River (Figure 1.6). At about 1500h on September 21, local settler Frank Gordon saw smoke west of his farm. He assumed settlers were burning to clear land and therefore did not check the source of the smoke. The following day, LaFoy noticed the same smoke coming from the west and investigated its source. When he arrived at 1700h, the strong southwest wind already swept the fire across the muskeg and willow and into Gordon's farm. LaFoy dispatched two bulldozers and moved his locally hired firefighters to this fire. The settler who leased the land where the Outlaw Fire started had earlier cleared his land using fire but was reportedly away for two weeks before the fire blew up. In the fire report, LaFoy stated he believed embers from the big fires along the British Columbia–Alberta border started this fire.[85]

The firefighters managed to control the Outlaw Fire on September 22 but moved again three days later, when another fire called the Naylor Hills Fire came down the south side of the Keg River and devoured Frank Jackson's sawmill (Figure 1.6).[86] The fire report completed by LaFoy describes the cause of this 874 ha fire as "Believed from wind carried embers from the interior fires 15–25 miles northwest."[87] Local farmer Harry Wilson saw fire on the Naylor Hills in the afternoon on September 25. Later in the evening, he rode out to check on the fire. After finding about 160 ha on fire, he rode back and notified LaFoy. When Jackson's sawmill burned, LaFoy received permission to hire a firefighting crew. The crew of 20 local trappers and farmers took about six hours to reach the fire. The *Naylor Hills Fire Report* stated that Jackson lost his sawmill but no lumber. Jackson, however, claimed the fire additionally destroyed his logs and 4,572 m (15,000 feet) of lumber.[88]

The fire report states, "The fire was burning in heavy sw [white spruce] timber and a part of it had gone by and over Blueberry Fire and on to the Highway by Willow Swamp Fire as it was being driven by very high Southwest winds on this date."[89] E. Strong, the manager of the Hudson's Bay Company trading post spotted the Blueberry Fire at 1230h on August 28, in an area where approximately 150 people harvested berries on Sunday, the previous day. The fire originated from an abandoned campfire started by one of the blueberry pickers. When Strong arrived, the 0.4 ha fire was burning in heavy windfall and jack pine; unable to stop it, he rode back and telephoned LaFoy. The Blueberry Fire increased to 20 ha when strong southwest winds associated with a frontal passage occurred. One bulldozer and a crew of 12 firefighters arrived at 1800h, but rain just behind the cold front arrived at the fire 30 minutes earlier. With a fireguard established the next day and a little more rain two days later, LaFoy sent the crew home. He declared the Blueberry Fire extinguished on September 11, the same day he commenced firefighting on another fire called the Willow Swamp Fire.

Campers along the Mackenzie Highway caused the 0.4 ha Willow Swamp Fire.[90] LaFoy discovered this fire at around supper time on September 10. He tried to extinguish the fire but the winds forced him to retreat and seek help. Six trappers arrived at 0600h the following morning and started to build guard. On September 13, the guard completely circled the fire, thereby containing any further spread. The fire nevertheless continued to burn in the 0.6 m to 0.9 m layer of moss and organic matter. The next day, 32 km/h southwest winds blew the fire across the guard. The two patrolmen left behind to watch the ground fires worked hard without a water supply to stop the excursion. Everything was under control until September 25 when the Naylor Hills Fire struck. It spread south and ran into both the Blueberry Fire and the Willow Swamp Fire.

The fires LaFoy chased became lost at the provincial level; too many other fires took precedence. Over 30 fires continued burning in northern Alberta on Friday, the hottest September 22 on record.[91] Only two of these fires were under control. The number of men engaged in fighting fires in the province then totalled 700.

Settlers continued fighting the Wanham and Newbrook and another fire near Fishing Lake, northeast of Edmonton. The two priority fires for the Alberta Forest Service were the Lac La Biche Fire and the Iosegun Fire (northwest of Whitecourt). The Alberta Forest Service managed to hold the Lac La Biche Fire from burning into merchantable timber. Twelve men and one bulldozer were dispatched to fight the Iosegun Lake Fire. The bulldozer was ferried across the Athabasca River on a raft because of the difficult access to the fire. Alberta was winning some important battles but not the war. As fires were brought under control, new fires arrived. From September 15 to 18, ten new fires started. Officials finally acknowledged it was the worst fall fire season in 20 years. The alarm finally sounded.

An additional seven bulldozers and two watertank trucks arrived on the Wanham Fire.[92] The province borrowed these resources from a contractor who was clearing land 40 km from the fire. The bulldozer is a firefighter's friend in the boreal forest. A good operator can efficiently build many miles of fireguard by removing the vegetation or covering it with clay. The borrowed bulldozers arrived at the Wanham Fire at 0300h on Sunday, September 24. The sound of the large machines building guard brought hope to the exhausted settlers who were still fighting the fire.

A front-page article in the *Edmonton Bulletin* on Monday, September 25 stated "100 Bush Fires Ravaging North."[93] In this article, Forest Service officials in Alberta estimate they have 44 active fires. Jack Jaworski, flight lieutenant with the Royal Canadian Air Force, gave a riveting account of the amount of fire on the landscape during his flight from Whitehorse to Fort Nelson and then to Moberly Lake, British Columbia:

> On the flight down, we saw four huge major fires stretching from northeast of Fort Nelson [British Columbia] to a point near Grande

Prairie [Alberta]. They were raging in a stretch over 300 miles east of the Mackenzie Highway. Besides these, there were a number of smaller fires, west of the highway. And from a point near Peace River, Alta., in a 200 mile long line to the Rocky Mountains, there was an almost continuous fire front. South and west of Dawson Creek, BC we counted 50 separate fires covering another huge area. The whole body of fire and smoke seemed to be moving to the northeast before the wind. It looked like a whole big chunk of the northland was on fire.[94]

Jaworski concluded, "All they can do is pray for rain."[95] He estimated it would take three days of continuous rain to stop the fires. Even when the rain did arrive, two of the larger fires—Blueberry Fire and Beatton River Fire—continued to burn on Thursday, September 28.

An article in the September 26, 1950 issue of the *Globe and Mail* also described Jaworski's flight over the fires. This story reported that Jaworski counted 60 fires in northeastern British Columbia and another 40 fires burning out of control in northwestern Alberta. The four largest fires occurred along the Mackenzie Highway in a similar location as described in the *Edmonton Bulletin* article. Jaworski estimated the largest fire to be 190 km wide by 200 km to 240 km long.[96] This equates to an area of 3.8 million hectares to 4.56 million hectares. Although likely an overestimate, it suggests earlier estimates of area burned were underestimated.

The observation made by Jaworski indicates two large areas of the boreal forest burned during September in 1950—a 480 km band along the Mackenzie Highway in northeastern British Columbia, and a 320 km band in Alberta from Peace River west to the British Columbia–Alberta border. The historical weather data supports Jaworski's account of the two worst days occurring on September 20 and 22.

Table 1.1 lists the known active fires greater than 200 ha from September 20 to 22, 1950 in northeastern British Columbia and northwestern Alberta. More fires occurred than were officially reported. There were just too many fires and not enough suppression resources to action them all. Access was also a problem. When the smoke cleared, an estimated 2 million hectares of the boreal forest burned. This is one of the largest firestorms ever documented (Figures 0.2 and 1.7).

On September 23, the fires along the Mackenzie Highway disrupted the Northwest Communications System that provided civilian and military communication.[97] At mile 255 of the Mackenzie Highway, the fires destroyed 280 telegraph poles along three kilometres of the Canadian National Telegraph Line. This severed the only direct land communication line to the State of Alaska. Restoring important military communications, including airport control for the northern areas, quickly became a priority. Crews worked 24 hours to restore communications using a temporary line.

On September 24, forest rangers reported five new fires.[98] Eighteen firefighters and one bulldozer were dispatched to fight two fires near Buck Lake. Another 21 firefighters fought a fire in the Lac La Biche district. Two other fires started in the Whitecourt and Drayton Valley areas.

The only structures lost when embers rained on the Keg River prairie was a barn with a thatched roof, Gordon's mill and grain bins, and Jackson's sawmill. The other buildings remained unscathed due to the efforts of the homesteaders and trappers who stayed to protect their properties and the recently harvested crops. Louis Jackson used his bulldozer to pull houses into the safety of the harvested fields. It was a community effort. When rain and snow fell on October 4, a tired LaFoy welcomed the chance to get some rest. The next morning, LaFoy looked outside and, for the first time in three weeks, saw the sun.

Dragging homes away from a fire's reach is not an option today, and there is considerable debate whether homeowners should stay to protect their property or evacuate during a firestorm. Until 2009, homeowners in Australia were advised to stay and defend their homes if their properties were properly prepared for a bushfire. Most houses are destroyed after the front of the fire has passed. The decision to stay or evacuate must be made early. Many homeowners die because they decide to evacuate too late and the roads are already blocked because of fire and smoke. They didn't know whether the exit route they chose was safe. Wildfires in the boreal forest are not the same as bushfires in Australia, but the need to prepare for wildfire is the same. People who live and work in a flammable forest must plan and prepare for wildfire to reduce the impact on their property.

TABLE 1.1 Fires greater than 200 ha, northeastern British Columbia and northwestern Alberta, September 20–22, 1950

Fire Number	Fire Name	Province	Size (ha)	Cause	Duration
–	Chinchaga River (No fire report)	AB	988,497	Human	?–Oct. 4
133	Chinchaga River	BC	123,491	Human	July 29–Nov. ?
19	Whisp (Chinchaga River)	BC	147,705	Human	June 1–Oct. 4
–	Chinchaga River (No fire report)	BC	202,458	Human	June 1–Oct. 4
848	Goose Mountain 4	AB	346	Lightning	Sept. 13–27
849	Hogs Back	AB	3,580	Human	?
850	Wildwind	AB	768	Human	?
851	Clear Hill	AB	449	Human	Sept. 12–23
852	School	AB	616	Human	Sept. 17–26
853	Traders	AB	1,152	Human	Sept. 17–21
854	North	AB	834	Human	Sept. 23–26
855	Rat	AB	332	Human	Sept. 11–Oct. 10
856	Naylor Hills	AB	874	Human	Sept. 25–Oct. 19
857	Outlaw	AB	290	Human	Sept. 21–25
864	Puskwaskau 30	AB	1,683	Human	Aug. 30–Sept. 27
865	Puskwaskau 35	AB	974	Human	Sept. 12–Oct. 12
866	Puskwaskau 41	AB	1,894	Human	Sept. 18–Oct. 18
867	Puskwaskau 45	AB	1,890	Human	Sept. 22–Oct. 5
868	Nipisi	AB	1,262	Human	Sept. 17–Oct. 13
873	Wanham	AB	4,330	Lightning	Sept. 14–25
877	House River 22	AB	35,985	Human	Sept. 14–Oct. 14
878	Duck Hill	AB	320	Human	Sept. 13–22
879	No known fire name	AB	1,024	Unknown	?
880	Sask. 36 Blueberry Mountain	AB	18,432	Human	Sept. 9–25
113	No known fire name	BC	122,000	Unknown	July 29–Nov. ?
118	No known fire name	BC	807	Unknown	July 29–Oct. ?
146	Bess	BC	2,930	Human	Sept. 20–Oct. 18
147	Brad	BC	540	Human	Sept. 20–26
148	Duke	BC	939	Human	Sept. 21–26
151	No known fire name	BC	25,021	Unknown	Sept. 20–Oct. ?
152	No known fire name	BC	1,488	Unknown	Sept. 20–?
153	No known fire name	BC	6,457	Unknown	Sept. 22–Nov. ?
165	No known fire name	BC	7,910.	Unknown	Sept. 21– Nov. ?

As a result of the horrendous fires that burned in the Australian State of Victoria in early February 2009, a royal commission investigated the "stay and defend" strategy versus the "leave early" strategy. The commission concluded that better information and warnings are required to inform homeowners whether they should stay or go. The fires killed 173 people—Australia's worst loss of life from bushfires—and a total of 2,133 homes were destroyed.[99] When the Black Saturday bushfires ripped through the scenic town of Marysville, only a handful of buildings remained standing. The next morning, ABC reporter Jane Cowan visited the remains of Marysville and described the scene: "We were in the main street and it's like a warzone, like a bomb has been dropped on the entire township," she said.[100]

The 7,500 people left homeless from the 2009 fires in Australia received assistance from their government and insurance policies. The residents at Keg River were not so fortunate. With no insurance and no money to rebuild, the Keg River homesteaders had no choice but to stay and fight the fire. The large burning spruce cones flying through the air ahead of the fire rained fire on the settlement. Everyone participated to extinguish the many spot fires, including small children who hauled buckets of water from the Keg River. Remarkably, no one died. Trapper Larone Ferguson narrowly escaped the fire while working his trapline just west of Keg River.[101] Unable to keep ahead of the fire, the river became his only available escape route. His horse refused to follow him down the bank and into the river. With badly skinned and bruised hands, Larone let go of the rope and jumped into the Haro (Big Creek) River, keeping only his eyes and nose above water. He survived the fire, but burned his eyebrows and hair when a wall of flame roared overhead across the river. His horse disappeared.

The Jacksons

Frank Jackson's wife, Mary Percy Jackson, was the local doctor in Keg River. In 1929, the Government of Alberta placed an advertisement in the *British Medical Journal* specifically seeking strong, energetic women doctors with obstetrics experience to work in northern Alberta. The ability to ride a horse was considered an asset for this job. Male doctors were not encouraged to apply because they usually abandoned their northern posts after a short stay to work in larger communities in the south. Mary Percy won the Queen's Prize in 1927 as the top medical student in her graduating class at the University of Birmingham, England. After working two years at Birmingham Children's Hospital, she decided to apply for work in obstetrics in Calcutta, India. Since the next post for this position did not open until the following year, she decided first to seek adventure in western Canada and applied for the job in Alberta. Her multiple degrees, including a bachelor of medicine, bachelor of surgery, and two medical diplomas, combined with her experience working as a physician and surgeon, and a love of horses, meant that she easily won the competition. In June, Mary set sail for Canada. Her planned one-year assignment would turn into a dedicated life of service to Albertans living in the Peace River region.[102]

After travelling by steamship, train, paddle wheeler, and horse and wagon, Mary Percy arrived at Battle River, now the town of Manning (renamed in honour of E.C. Manning, the premier of Alberta from 1943 to 1968), located 160 km south of Keg River. From her small, five-metre by six-metre one-room house along the Notikewin River, Dr. Percy immediately started practicing medicine. She was often called upon to travel by horse to visit patients who were too sick to travel to Battle River. Her indoor riding lessons at the Sutton Coldfield Riding School in England were quickly put to test when the locals collected 31 dollars and bought her a horse.[103] "Mary Percy once rode 51 km a day to visit a seriously ill patient until the road, wet from rain, dried enough to allow for his transport by vehicle to the hospital in Peace River."[104] On these long rides, Mary Percy's moose-hide clothing was more suitable in the wilds than the formal riding clothes she brought with her from Britain. She described the clothes as "delightfully warm and yet fairly lightweight."[105]

Completed in 1948, the Mackenzie Highway extends north 465 km from the Town of Grimshaw to the Alberta–Northwest Territories border. Construction started briefly before the Second World War. After abandoning the project during the war, work resumed as a postwar project funded by both the dominion and provincial governments. In the early days, the Peace River was the lifeline to the small settlements along or close to the river. During the summer, riverboats carried freight and passengers between Peace River, Fort Vermilion, and Fort Smith. The river set the pace—its water levels dictating how much cargo could be carried and when it would arrive. During the winter, smaller loads were pulled by dog sled on pack trails or on the frozen river.

Carcajou Point (formerly known as Wolverine Point) was the first settlement in the area, located on the east bank of the Peace River about 240 km downstream from the community of Peace River. When J.W. Pierce surveyed Carcajou Point in 1916, it was a small settlement of Native and Metis hunters and trappers.[106] The nearby prairies to the east provided feed for the few cattle and horses they kept.

In 1908, Revillon Frères opened a trading post in Keg River to help stock their fur stores in Paris.[107] The next year, the Hudson's Bay Company arrived in Keg River and built a trading post and store nearby to compete with their French rivals.[108] Revillon Frères closed operations at Keg River during the Depression in the 1930s. Locals such as Frank Jackson also entered the fierce fur trade war. He successfully ran the Keg River Trading Company from 1921 until 1949.[109] The Hudson's Bay Company operated the only trading post and store to remain open in Keg River until 1970.[110]

The riverboats delivered supplies for the trading posts at Keg River. Once unloaded at Carcajou, the supplies were transported across the river and stored in warehouses on the west bank of the river. Weather permitting, horse and wagon hauled the supplies down the Keg River Trail to the trading posts.

When Frank Jackson lost 50 cows during the severe winter of 1919–20 at Red Deer Lake, he decided it was time to move. In May 1920, he took a scouting trip to the Keg River prairie.[111] Despite being warned about the black flies, mosquitoes, horse flies, and sand flies, Jackson returned to Red Deer Lake and proudly announced to his wife, Louis, that Keg River would be their new home. Biting insects did not cause any problems for Jackson when he first visited the Keg River prairie in May, but they were waiting for him on his return trip. When Jackson disembarked from the paddle wheeler steamboat docked at Keg River Landing across from Carcajou on the Peace River, he received a boreal forest welcome from the ferocious flying bloodsuckers. This initiation to the boreal forest, and other hazards he later encountered, did not deter Jackson. He was a stubborn man, determined to realize his dream of establishing a successful homestead in Keg River. Jackson succeeded. When land in the Keg River District opened for homesteading, he used an 18-inch walking plough and quickly broke his newly obtained land. His first harvest of about one hectare of oats also became the first harvest for the district. In 1986, the development of a new variety of barley called "Jackson" commemorated Frank's contributions to establish an agriculture presence in the Keg River area.[112] This was a hardy variety intended to thrive in the north, just as Jackson did.

Dr. Percy became the first woman doctor in northwest Alberta, and the only doctor between Peace River and Fort Vermilion.[113] She planned to stay one year and then return home to England. These plans changed in January 1930, when Frank Jackson travelled 160 km from Keg River to Battle River by dog team, in -40 °C to seek treatment from Dr. Percy for a serious hand infection.[114]

Frank's visits to Battle River continued, and on March 10, 1931 he married the woman who not only healed his hand but also won his heart. The previous year was difficult for Frank. The Depression hit, causing fur prices to spiral downward. His first wife, Louise, then became sick after giving birth to their third son. Louise needed medical attention urgently, but without phones or radio communication, the only way to dispatch a medical evacuation flight to Edmonton was to send a messenger on horseback to Peace River. The plane arrived, but it was too late. She died soon after being admitted to the hospital in Edmonton.

Dr. Mary Jackson moved to Frank's homestead at Keg River and continued practicing medicine until her retirement in 1974.[115] Frank was her last patient. He died in 1979 at the age of 87. Frank and Dr. Percy had much in common. They enjoyed the outdoors, music, and hard work. They also shared a common heritage. Frank was born in Southend, England in 1892.[116] He moved to Canada with his father in 1904. Dr. Jackson lived until 2000. She was 96 years old.[117]

It took Dr. Mary Jackson over four weeks to travel from Birmingham, England to Battle River, Alberta. The smoke from the Chinchaga Firestorm took only four days to make the return trip.

Fires, Floods, and Rabies

Besides the destruction of the forest and the endangerment of people's property and livelihoods, the Chinchaga Firestorm had often overlooked effects. For example, the fire displaced many animals, which itself had several consequences. The fire also reduced soil stability, which led to flooding the following spring.

Dr. Mary Jackson remembered the fire well. In a 1984 letter to Peter Murphy, she wrote, "The fire came in from the northwest, and there were bears and moose and deer all coming together, not bothering with each other, but running before the fire. During the day, the sky was dark as night, and at night the sky was lit up like sunset all around the horizon."[118] Many animals were reportedly killed by the fire or from the effects after the fire. Dr. Mary Jackson's neighbour, Dave MacDonald, flew over the area after the fire and reported seeing many moose carcasses.[119]

Professor William Rowan, head of the Department of Zoology at the University of Alberta, investigated the effects of the fire on wildlife populations. He observed many dead moose carcasses along the banks of the First Battle (Notikewin) River; the moose likely died because of high levels of sediments, nutrients, and metals that contaminated the water.[120] LaFoy also found many dead animals, particularly moose, along the banks of the same river.

The Chinchaga Firestorm displaced many animals, which resulting in higher densities of animals north of the burned area. These higher densities may have contributed to the rapid spread of rabies in 1952 in the north Peace River region. The 1950 fire and the 1952 rabies epidemic significantly reduced wildlife populations in the Keg River area. Despite the high number of animals killed, their populations rebounded and, according to LaFoy, trappers returned after about 25 years.[121]

Fearing that the rabies discovered in 1952 in red foxes in northern Alberta would spread, the provincial government embarked on a massive poisoning campaign from 1952 to 1956. The rabies disease in the red fox population in northern Alberta was enzootic—that is, present but only manifesting itself in a few cases. The disease began to spread when the number of red foxes reached their apex. In 1953, the disease spread to coyotes. It then spread from coyotes to domestic animals. Concerned about the threat of decimating Alberta's cattle industry, politicians endorsed a war against rabies. Over two million strychnine baits later, the program of extermination recorded killing 5,461 wolves, 50,781 coyotes, and 55,499 foxes from forested areas, mainly in northern Alberta. An estimated 100,000 to 120,000 additional coyotes were killed in agricultural areas.[122] The poisoning campaign also impacted non-target predators and scavengers. The poisoned bait killed indiscriminately. The death toll also included fisher, marten, lynx, cougar, and bear.

In 1951, the Chinchaga River flooded.[123] Fires and floods form a dangerous liaison—one disaster seeking another. When severely burned areas in mountainous or hilly terrain experience high precipitation during the following winter or spring, flooding inevitably occurs. The Chinchaga River Fire burned the vegetation cover on the slopes of Halverson Ridge, which reduced the soil stability and resulted in erosion and flooding.

In June 1897, extensive flooding occurred in the eastern Rockies. William Pearce, inspector of irrigation, wrote in 1898, "The disastrous floods which were experienced last June were chiefly, if not wholly, the outcome of the destruction of the forests along the foothills and eastern slopes of the Rocky Mountains....Last season's experiences will be repeated if steps are not taken to preserve the trees and brush at present growing on the areas mentioned."[124]

The attraction between fires and floods still occurs today. On Christmas Day, 2003, heavy rains in California triggered flash floods and mudslides in an area of the San Bernardino Mountains, where two months earlier a high severity wildfire called the "Old" fire burned through the upper Waterman Canyon. A sudden downpour of 8.9 cm (3.5 inches) of rain caused flash floods and mudslides that smashed through a youth camp. Fourteen people, most of them children, died.

The 1951 Chinchaga River flood was uneventful. No one died and no losses were documented in the annual report for the Department of Lands and Forests. Like many of the ghost fires that occurred, the 1951 flood received little notice.

In remote areas of Canada's boreal forest, small fires still occur without being mapped and recorded. Large fires, however, cannot escape the eyes of the many satellites orbiting the earth. Contemporary forest rangers are also equipped with a suite of tools, including helicopters, thermal infrared cameras, global positioning systems, geographic information systems, satellite data communication, and detailed spatial data about the fire environment.

The twenty-first-century world is a more complex and uncertain world than LaFoy experienced. Nevertheless, the challenges LaFoy encountered would challenge many forest rangers today. When the Hudson's Bay trading post ran out of food, LaFoy ordered supplies from Battle River. He was the logistics section chief responsible for all services and support for fire incidents. Without a Forest Service vehicle, LaFoy used his personal truck. When it broke down, he became the mechanic. LaFoy was responsible for food, communications, medical needs, supplies, and camp setup. He did not have a helitack-rappel crew and helicopter based in Keg River waiting for fires to arrive. LaFoy had to find able working men. This meant first getting permission from the Peace River Division Office and then phoning and driving around the Keg River prairie, all while the fires continued to grow.

When fires arrived, LaFoy had to assess the situation and determine what resources were required. More importantly, he needed to plan a strategy and tactics. He was thus the planning section chief. As the person responsible for managing the fires, he too was the operations section chief and the incident commander; sometimes he was both incident commander and the firefighter, building the guard with hand tools. He was also the safety officer, information officer, fire investigator, and documentation team lead. In 1950, the forest ranger did it all. No one welcomed the rain and snow on October 4 more than LaFoy. He saw more smoke and fire than forest rangers in the south would see during their entire careers.

In the spring 1951, LaFoy moved out of the Peace River Division to High Prairie, located just west of Lesser Slave Lake. He decided to leave forestry and move to Rimbey to start a new career with the Game Branch in 1953. There he shone until his retirement in 1970. In 1986, LaFoy received the distinguished Alberta Order of the Bighorn Award in recognition of his exceptional contributions to fish and wildlife conservation in Alberta. Frank "Trapper" LaFoy continued to run traplines and teach others how to trap until 1993. He passed away on January 28, 1995.

TWO

BLACK SUNDAY

On June 17, 2000, Betty Rhodes posted a query under "Erie Legends" on the *GoErie* website, a source for local Erie, Pennsylvania news and events, regarding a mysterious "Black Sunday" in 1950. She wrote,

> One Sunday afternoon the sky got really dark—dark enough to view the stars. The skies stayed dark for the rest of the day. We did not own a TV set at that time, and with the whole family wondering what was going on (some thought the end of the world was at hand), my dad took us to Uncle Jake's in Corry to see what his TV set had to say about the situation. After many hours of no announcements or bulletins about the particular event, finally a reading

came on the bottom of the screen that said there was a big forest fire in Canada and this was causing the darkened skies.[1]

When Black Sunday occurred, Betty was an eight-year-old girl living in Wattsburg, Erie County, Pennsylvania. She hoped others who experienced this day might answer some of the questions that still puzzled her: "If a forest fire was sending smoke to darken the skies why could we see the stars? And why doesn't the sky darken on that magnitude with the forest fires today in our country?"[2] She also wanted to know the date and location of the fire.

Betty received many responses to her query. One person replied that she, too, was a little girl on Black Sunday, September 24, 1950. Although she couldn't provide any details, she clearly remembered her family being very frightened. They thought it was the end of the world. She was not the only person who thought doomsday arrived. That Sunday, a minister in Ashtabula, Ohio spoke about the coming of the end of the world during his sermon. He did not tell his congregation when the end would come, only that it would come "in an orange glow."[3] As the parishioners drove home after the sermon, the sky darkened with an eerie yellow glow. By mid-afternoon, the darkness spread southward from Illinois to Pennsylvania. In Bradford, Pennsylvania, day turned to night at about 1430h. During the two-hour blackout, the Bell Telephone Company received an estimated 400 to 500 calls at their Bradford office alone.

Henry Carlson lived alone in a small shack in Jamestown, New York. He lived without a radio, television, or phone. He was trembling when his nephew came by to check on him. "Do you think this is the end of the world?" Carlson asked his nephew.[4] When darkness suddenly falls during the day, chaos, panic, and fear ensue. On September 24, 1950, the rumours spread as fast as the fire itself. Police stations and newspaper offices were inundated with telephone calls. Was it a tornado, snowstorm, or maybe a poisonous cloud of smog? Some people thought this strange phenomenon was the result of a nuclear explosion or a secret military weapon test. Others believed it was an invasion of flying saucers. In 1950, flying saucers and secret military experiments were high on the public's fascination list. Even today, some people who experienced that day when the sun went out still question what caused the blackout. Perhaps a tired sun was loosing its energy. This seemed plausible for some callers who asked if, indeed, the sun was going out. Residents in Tillsonburg, Ontario had another theory: a greater power's annoyance with the mortals on earth for tampering with standard and daylight times.

A panic stricken woman in Hamilton, Ontario phoned the United Press Hamilton Bureau to report, "I've just received a long-distance call from Cleveland. A radio station there has announced that Canada is on fire."[5] Many respondents to Betty's query indicated they heard Black Sunday was the result of a large fire in Canada, but they didn't believe it. How could so much smoke travel so far and cause so much darkness? And why couldn't they smell the smoke? It must be the Central Intelligence Agency (CIA), some of the respondents concluded. Despite confirmation of the forest fires in western Canada as the source of the smoke, media reports that suggested otherwise continued to circulate; they offered volcanoes in Japan, coal dust from Germany, and ice crystals in the upper atmosphere as explanations for the darkness. William Miller, a retired colonel with the United States Air Force, believed it was the result of the government experimenting with the use of powdered silver nitrate to seed clouds and subsequently provide cover for bombers on their missions. Miller's uncle, who worked as a pilot in the United States Air Force, claimed a seed cloud floated across the country from the desert in New Mexico where these tests were being conducted.[6] Miller feared the government's capability to cover up events such as Black Sunday. The rumour he heard about an accidental explosion of a nuclear bomb causing the darkness could not be discounted.

The cloud seeding explanation had some merit. Extensive cloud seeding experiments did, in fact, occur in California, Nevada, and New Mexico from 1945 to 1955. In 1946, Vincent Schaefer, a scientist with General Electric, flew into the top of a cloud and released three pounds of dry ice pellets.[7] The snow that fell five minutes later shifted Schaefer's rainmaking career into high gear. Despite limited success and increasing criticism, military funding continued for cloud seeding research.

From 1948 to 1950, the United States Navy and Air Force conducted many cloud seeding sorties.[8] Manipulating weather was, and still is, considered the ultimate military wildcard for success. If you can control

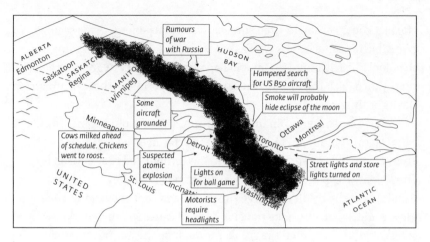

Rumours
of war
with Russia

HUDSON
BAY

Hampered search
for US B50 aircraft

Smoke will probably
hide eclipse of the moon

Some
aircraft
grounded

Cows milked ahead
of schedule. Chickens
went to roost.

Suspected
atomic
explosion

Lights on
for ball game

Street lights and store
lights turned on

Motorists
require
headlights

ALBERTA
Edmonton
Saskatoon
SASKATC
Regina
MANITO
Winnipeg
Minneapo
Detroit
St. Louis
Cincina
Washingto
Toronto
Ottawa
Montreal

UNITED
STATES

ATLANTIC
OCEAN

FIGURE 2.1   Rumours and events that followed the smoke pall.

weather patterns, you can control wars. Numerous cloud seeding operations occurred around the world in the 1950s in an effort to increase rain-
fall in drought areas. The military ran all of these operations.

One of the most popular theories suggested the US Department of
Defense was testing large smoke screens to use as a weapon during war,
similar to cloud seeding. Producing cover to hide bombers made sense. If
you cannot see the enemy, you cannot kill the enemy.

Black Sunday created a lifelong impression for Larry Knickerbocker.
He was five or six years old living on West 7th Street in Erie,
Pennsylvania when the skies darkened.[9] His dad confirmed it happened on a Sunday afternoon, and they both agreed the sky cleared by
nightfall. Larry remembers the sky to the north being very distinct with
traces of semi-transparent, smoky looking clouds. Larry's father, like
many other fathers who survived a war, questioned the credibility of the
Canadian forest fire story. He, too, implicated the military, convinced
the US Department of Defense was testing a secret weapon. This explanation was credible for many because on January 31, 1950, President
Truman had announced that the United States planned to proceed with
the development of a hydrogen bomb. These fears were captured in an
article on the front page of the *New York Times* on Monday, September 25,
1950: "Forest Fires Cast Pall on Northeast. Canadian [Smoke] Drift 600

Miles Long Darkens Wide Areas and Arouses 'Atom' Fears."[10] The 966 km long smoke plume was a reported 322 km wide.[11]

The residents of northern Ontario had good reason to believe the smoke came from an atomic bomb. Overhead, from September 21 to 23, the sky filled with military planes flying to thwart what many people believed was a Russian invasion. The Soviet Union successfully detonated an atomic bomb on August 29, 1949, which sent shock waves around the world and fear of an impending nuclear confrontation between the Russians and Americans. Unknown to the northern Ontario residents, the Royal Canadian Air Force was searching for a missing United States Air Force B-50 bomber. This giant plane crashed in Labrador while on a flight from Goose Bay, Newfoundland to Tucson, Arizona. All 16 personnel on board survived.[12] News of this story arrived too late for the Ontarians. They believed the invasion was well under way.[13]

In southern Ontario, the sudden darkness created an unexpected surge in demand for electricity. Ontario Hydro reported a record usage of 180,000 kWh to turn on the lights during midday. Some power lines subsequently failed. In Toronto, power line breaks set off bank alarms. Police racing to the banks in the dark with their car lights and sirens on only fuelled the fear and speculation surrounding this peculiar phenomenon. The September 25, 1950 *Toronto Daily Star* mapped the reported fears and events that followed the smoke pall (Figure 2.1).

The smoke did provide some reprieve for farmers in Ontario. A potential crop-damaging frost forecasted on Sunday morning did not occur because the layer of smoke acted as a thermal blanket trapping the long wave radiation emitted from the earth at night.

Cliff Shilling encountered the darkening of the skies while flying his Piper Cub airplane south from Kearsage to Brookville, Pennsylvania on September 24, 1950.[14] As he looked back northward, he saw a large dark area. Since no impending storms were forecasted, he circled several times to get a better look at this strange sight. As he continued flying south, the dark area closed in on him. Schilling's flight took one hour. Just before landing he noticed the smell of wood burning smoke in the cockpit. Fifteen minutes after landing, the sky became pitch dark.

Other pilots also reported encountering the dark band of smoke. About 48 km northwest of Erie, Pennsylvania, a Trans World Airlines pilot advised air traffic control he was flying by instrument through smoke at a height of 7,620 m (25,000 feet). He noted that above the smoke layer it was clear and sunny.

Darkness blanketed New York City between 1200h and 1300h on Black Sunday.[15] The sudden shift from day to night fooled Dora Gesaman's chickens in Watts Flat, New York. They all scurried back to the barn to roost. When the darkness lifted at 1600h, her rooster started crowing to signal the assumed start of a new day.[16] Agitated cows also needed to be milked ahead of schedule.

The smoke also arrived in time to dampen the celebrations of many groups in northeastern United States waiting for the lunar eclipse to start at 0920h on September 25. The smoke blacked out the eclipse and left a television show in New York that was trying to cover the event with little to report.[17]

The darkness fooled more than just the chickens and cows. A concerned London, Ontario woman phoned a radio station after she awoke to darkness, convinced she slept the entire day.[18] Some people rushed to work with their car lights on believing they were late for work. Wallace Barlow from Sugar Grove, New York took a nap after lunch on that day. When he awoke it was totally dark. Barlow, too, thought he overslept and missed his chores, but the kitchen clock said otherwise. He drove to the barn and met his father who was standing outside with a puzzled look.[19] Frustrated by the lack of information on the radio, many people living in rural communities learned about the fires in Canada by resorting to their last line of communication—the party telephone line.

Even in Pittsburgh, Pennsylvania where the sun never shone because of air pollution from the mills operating continuously to produce iron, steel, and aluminum to feed the war effort, Black Sunday was an event to be remembered. Jayne Mortenson, who was ten years old at the time, described it as "the summer of the orange skies and strange atmospheric effects."[20]

Most of the fires burning in northeastern British Columbia and northwestern Alberta were free-burning fires. Some of these fires likely

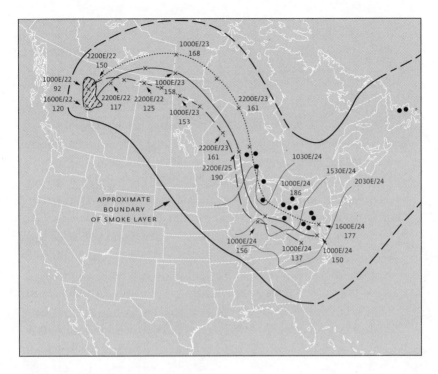

1000E/22, 92   10:00 am; September 22, 1950; 9,200 feet

●   Weather stations that reported a significant reduction in daylight on September 24

———   Approximate boundary of smoke layer

————   Isochrones at five-hour intervals

·····×····   312 K surface isentropic forward trajectory of the smoke from the fire area starting at 2200h
September 22 at a height of 4,755 m (15,600 feet)
— ✳ —   312 K surface isentropic forward trajectory for the smoke from the fire area starting at 1000h
September 22 at a height of 2,804 m (9,200 feet)
——×——   312 K surface isentropic backward trajectory from Washington, DC at 1000h on September 24
at a height of 4,572 m (15,000 feet)

FIGURE 2.2   Approximate trajectories and perimeter extent of the smoke pall. Reported observations of the smoke (time, day, and height) are noted along each trajectory. Eastern Standard Times (24 h) are used, and heights are shown as x100 feet. The isochrones show the southern movement of the smoke layer on September 24. The weather stations in Newfoundland reported darkness on September 25.

burned for several years, over-wintering as deep smoldering muskeg fires under the snow. Although British Columbia and Alberta generated the smoke that caused Black Sunday, they experienced the least impact from it. Most British Columbians and Albertans knew nothing of this

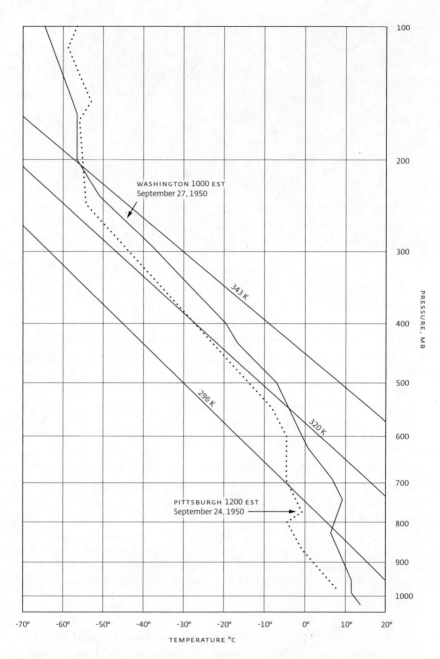

FIGURE 2.3 Upper air observations for Pittsburgh, PA at 1200h on September 24, 1950, and Washington, DC at 1000h on September 27, 1950. The three diagonals show potential temperature (isentropic) lines.

export until the smoke arrived duty free at Washington, DC at 1100h on September 24.[21]

By noon on September 26, observations of smoke and a blue moon were reported from the Isle of Man in the British Isles.[22] Other observations were made later that day and the following day in northern England, France, Germany, Spain, Portugal, Denmark, and Switzerland.[23] When the sun rose as an azure ball in Denmark, nervous citizens fled to their banks demanding their entire savings so they could flee from the unknown catastrophe they felt would surely follow this perplexing event.

Of all the fires burning, the Chinchaga River Fire was likely the single largest contributor of smoke in the atmosphere. In a three-day period from September 20 to 23, the fire spread an estimated 63 km.[24] On September 22, high winds of over 55 km/h and a low relative humidity (26%) resulted in very high fire intensities.[25] Since June 1, when the fire started, six major periods of fire spread occurred before the blowup on September 22.[26] As a result, when the fire took its last run, it had a fire front approximately 100 km wide. This large area produced a tremendous volume of smoke that eventually travelled half way around the northern hemisphere.

The United States Weather Bureau reconstructed the movement and extent of the smoke from September 24 to 30, 1950 (Figure 2.2). They collected reports of smoke from Canadian synoptic reports and sent questionnaires to 384 first-order weather stations operated by the bureau. Trajectories of the smoke path were also estimated using meteorological data, in particular, the 312 K (39 °C potential temperature) isentropic surface.[27] An isentropic surface is a surface of constant potential temperature. Air flows forward along isentropic surfaces. Since the 312 K isentropic surface is often not at a constant elevation, it can be used to estimate both the horizontal and vertical components of the flow of smoke. The known locations of the smoke observations were used by Harry Wexler as a tracer to check the estimated smoke trajectories. It was a good fit.

Clarence Smith from the Weather Bureau, Air Force and Navy (WBAN) Analysis Center was quick to publish the results and to describe the synoptic conditions in the September 1950 issue of the *Monthly Weather Review*.[28] Harry Wexler, chief of the United States Weather Bureau's Special Scientific

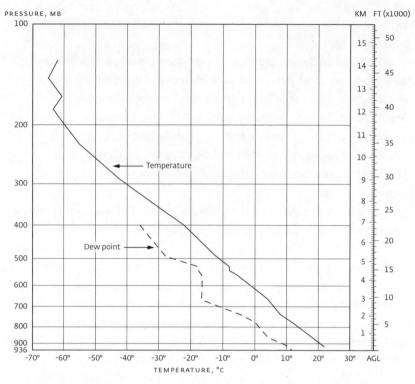

FIGURE 2.4   Radiosonde observations for Prince George, BC on September 22, 1950 at 1900h. Distances in kilometres (or feet) are shown as above ground level (AGL).

Services Division, published another paper titled "The Great Smoke Pall of September 24–30, 1950" in the December issue of *Weatherwise*.[29] Wexler used the same trajectory results as reported by Smith, but his estimate of the boundary of the smoke area is further south, to include observations of smoke across the State of Florida (Figure 2.2). Wexler also described the temperature and optical effects caused by the smoke.

No other smoke event from a single source rivals this event. The large volume, high concentration, and persistence of the smoke layer generated from the Chinchaga Firestorm distinguishes this event from other smoke events.

Special conditions were required for the smoke to obscure the sun and moon for such a long time and over such a large area. Black Sunday occurred because of the timing and location of air masses. A surface and upper ridge of high pressure persisted over western Canada from September 5 to 23. This drought period created ideal burning conditions once the high pressure ridge broke down and the winds arrived. Based on backward trajectories of the upper airflow along the 312 K (39 °C potential temperature) isentropic surface, Wexler estimated the smoke reached an estimated height of 3,840 m over the fire area.[30] Using the same trajectories, Smith estimated the smoke heights varied between 2,840 m and 4,755 m within the general area where the fires occurred.[31] At these heights, winds 103 km/h to 116 km/h carried the smoke in a northeasterly direction over northern Alberta and into the Northwest Territories until it hit the western edge of a deepening Artic low pressure system over Hudson Bay.[32] This cold air mass resulted in strong north-northeasterly wind flows just west of Hudson Bay. The smoke arched over Manitoba, Ontario, the Great Lakes Region, and north-central and eastern United States.

Weather stations in the Northwest Territories, northern Saskatchewan, and northern Manitoba first reported dense smoke on September 23. Once the smoke reached the Atlantic coast, it tracked northeast and returned back into Canada, over the Atlantic provinces, and continued across the Atlantic Ocean and over Eastern Europe. On September 24, a pilot reported flying through dense smoke at 5,182 m (17,000 feet) above Sault Ste. Marie, Michigan where the smoke report-edly first crossed the international border at 1030h.[33] Another pilot on a Trans-Canada Airlines flight to Toronto flew through smoke over the Great Lakes at a height of 4,267 m (14,000 feet). Upon landing, the crew noticed a film of oily residue covering the entire aircraft. This residue had obscured the pilot's visibility during his approach to land. He reported the smell of smoke in the cockpit and having to use the air-craft's windshield wipers to remove the residue.[34] The layer of smoke reached Toronto at 1345h and extended from about 2,134 m (7,000 feet) to 5,182 m (17,000 feet) above the city.

Other pilots reported the smoke layer at 4,267 m (14,000 feet) in the New York and Washington area. An American Airlines pilot flying from Cleveland, Ohio to Phillipsburg, Pennsylvania thought his plane had caught fire when he smelled smoke in the cockpit of his aircraft.[35]

As the smoke flowed quickly southward so did the cold air. On September 23 and 24, the Polar air mass moved southward into the northeastern United States. The cold dome of air created an inversion that persisted until September 29.[36] Generally, air temperatures near the earth's surface decrease with height. An inversion layer is a reverse layer of relatively warmer air above a relatively cooler layer below. Mixing is prevented because warm air rises and cool air sinks.

The inversion layer of cold Polar air prevented the smoke and warmer Pacific air from mixing and moving to the surface, which thereby prevented people from seeing and smelling the smoke. This also fooled some of the weather observers at the Washington Weather Bureau who mistakenly reported the smoke layer as cirrostratus clouds, cumulus clouds, and later as altostratus clouds. It wasn't until 1655h when the weather observers correctly reported the smoke and provided an official explanation of what caused the sudden darkness.[37] The inversion layer reached a height of approximately 4,267 m (14,000 feet or 600 mb) on September 24 (1200h) at Pittsburgh, Pennsylvania.[38] Three days later in Washington, DC, the top of the inversion layer was about 2,591 m (8,500 feet or 740 mb) in height (Figure 2.3).[39]

A second inversion layer located above 4,572 m (15,000 feet or 600 mb), although weaker than the lower inversion, prevented the smoke from moving further upward.[40] The two inversions created a dense smoke sandwich that blocked 50% to 60% of the sun's radiation. This reduced the forecasted maximum temperatures by 5 °C.[41]

The upper air sounding for Prince George, British Columbia on September 22, 1950 (1900h) shows two different inversion layers (Figure 2.4). The temperature at the surface was 15.8 °C. The first sounding recorded a temperature of 22 °C at a height of 325 m (1,066 feet or 915 mb) above ground level (AGL). This early evening inversion would only have a local effect of creating smoky conditions in the valleys. A second but weaker inversion occurred at about 5,500 m (18,045 feet or 500 mb). Above this height, the atmosphere became increasingly colder.

Smoke injected to different heights can travel in different directions. Estimating the height of the smoke from the 1950 forest fires in western Canada before it began its journey is difficult because there were many fires and hence many smoke columns and possible injection heights. The conditionally unstable atmosphere over western Canada suggests a parcel of air could have lifted to the tropopause at an estimated height of 12,070 m (39,600 feet). Other forces, however, were at play. Rising air (smoke) is also pushed by winds and, at lower elevations, channelled by topography. The energy of the fire, too, can influence the smoke's journey. Strong upper winds compete with the smoke's uplift. At a height of 5,510 m (18,077 feet), 74 km/h winds pushed the smoke laterally (Figure 2.4).

Flight Lieutenant Jack Jaworski, operations officer at Northwest Air Command in Edmonton, viewed the fire area while flying a Royal Canadian Air Force (RCAF) Dakota from Whitehorse to Edmonton on September 22. He left Whitehorse at 0600h with supplies and passengers on board. Two of the passengers were forced to fly to Whitehorse on an earlier northbound flight when the pilot could not land the plane at their destination of Fort Nelson. The RCAF responded to the call for help. In thick smoke, Jaworski managed to deliver the passengers safely. "We landed at Fort Nelson in a heavy smoke that made it impossible to find the landing strip. But we finally got in to unload our passengers," reported Jaworski.[42]

One hour earlier, Jaworski's commanding officer also landed at Fort Nelson. Two days earlier the commanding officer left Fort Nelson on September 20 for a test flight to Fort St. John. It was a harrowing trip. He encountered turbulence extending to a height of 4,267 m (14,000 feet). When the cockpit quickly filled with smoke, the commanding officer lost visibility of the instruments. Without supplemental oxygen, he would not have survived. Jaworski took his commanding officer's advice to avoid the fires and flew southwest to Moberly Lake, British Columbia, and then east back to Northwest Air Command in Edmonton.

Trapped in an anti-cyclonic eddy, smoke from the fires in western Canada lingered for almost a week over the eastern United States.[43] Some of the smoke continued to travel as far south as Florida, Georgia, and Tennessee. The smoke that did not get trapped in the eddy changed direction and headed northeast, reaching Newfoundland on September 25.[44] In Grand Bank, children attended school on September 26 with the classroom lights on.[45] On September 30, the smoke disappeared and clear skies returned to Washington, DC.[46]

THREE

DARK DAYS IN THE PAST

Smoke from other wildfires in the past also caused dark days. On May 19, 1780, dark clouds travelling an estimated 40 km/h spread south from Portland, Maine to Providence, Rhode Island.[1] Four hours of darkness descended on what became known as Black Friday. The wildfires brought more than daytime darkness for the early settlers; fear and panic spread quickly.

Daytime darkness is associated with noteworthy events. As Moses stretched his hands towards the heaven, a thick darkness fell over Egypt and lasted for three days.[2] At the sixth hour of the crucifixion, darkness blanketed the entire earth for three hours.[3] When Black Friday occurred on May 19, 1780, many people thought it was the second coming of Christ.

Written accounts and fire scar analysis indicate the fire season in 1780 was one of the worst before 1850 in northeastern North America.[4] Trees that survive a fire are often scarred. When a portion of a tree's cambium layer (living tissue) is killed by fire, it cannot grow again. The wound forms a scar; the year the injury occurred can be determined by counting the tree's annual rings from the current known date to the scar.

The fire scar dates suggest wildfires burned freely in 1780 in both Canada (northwestern Ontario and the Algonquin Highlands in eastern Ontario) and the northeastern United States (western Maryland, western Virginia, Missouri Ozark Highlands, Boston Mountains of Arkansas, and the Boundary Waters Canoe Area of northern Minnesota). Although most of these fires occurred in the United States, some of the smoke that caused the dark day on May 19, 1780 came from fires in Canada.

The northeastern corner of the United States became home for some of the first European settlers in 1620. Bounded by Canada, the State of New York and the Atlantic Ocean, this region, known as New England, experienced dark days on May 12, 1706, October 21, 1716, and August 9, 1732.[5] The only other documented dark days before 1786 in the United States or Canada occurred on October 19, 1762, when darkness over Detroit, Michigan,[6] and during October 1785, as intermittent periods of darkness fell across eastern Canada from New Brunswick to eastern Quebec.[7] Other dark days may have occurred as a result of wildfires but, without people in the area to observe these events, they were not reported and documented. Explorer Henry Youle Hind gave another explanation for the occurrence of dark days in 1785 in Lower Canada, but not in Upper Canada. He concluded only civilized and learned people populated Lower Canada. Their apparent absence from Upper Canada resulted in a lack of accounting of dark days. The importance of prevailing synoptic weather patterns were, at that time, not understood.[8]

Interestingly, the largest fire documented in eastern North America did not result in a reported dark day. On October 7, 1825, numerous settler and logging fires coalesced to form the Miramichi Fire. This devastating fire burned 864,000 ha in New Brunswick and 336,000 ha in Maine and killed a confirmed 160 people.[9] An estimated 280 to 300 lives were lost; the death toll is probably higher given that 3,000 lumberman worked in the Acadian forests to meet the strong demand for wood from Britain.[10]

The 1910 Great Idaho Fire in northern Idaho and western Montana burned over 1.2 million hectares in two days and killed 86 people.[11] On August 20, tornado-force winds merged many separate fires, creating a firestorm with an 80 km flame front.[12] Smoke from these fires travelled as far east as the St. Lawrence River. Dark days were recorded in the northern United States and southern Canada from August 20 to 25, as a large pall of smoke changed day to night.[13] The 1910 fire was notable not only for its size but also the extent and length of the daytime darkness it created. No other recorded episode of dark days before 1910 impacted as large an area as the Great Idaho Fire.

Many fires also burned in western Canada in 1910. Chronicler of cultural and environmental history of fire, Stephen J. Pyne reports that an estimated 1.46 million hectares burned in the Canadian Rockies that year.[14] In her undergraduate thesis, Amanda Dawn Annand maps the 1910 fires along the eastern slopes of the Canadian Rockies in Alberta. Her area burned estimates indicate the 1910 fires had a much larger impact in Alberta than previously reported.[15] Rick Arthur, retired wildfire prevention officer in southern Alberta, estimated 404,686 ha alone burned in the forest reserves in southern Alberta from Mill Creek north to Red Deer River.[16]

The 1950 Chinchaga Firestorm ranks as the largest boreal forest fire event documented in Canadian history. Across the border, the 1910 Great Idaho Fire ranks as the largest forest fire documented in the United States. Both firestorms danced with very strong winds to the same choreography—a preceding dry winter, early spring, extended period of summer drought, and an unprecedented fire load. An identical stage was set for both conflagrations.

In 1910, the northern Idaho and western Montana forested landscapes were under siege from walls of flames from over 1,700 fires.[17] There may have been 3,000 fires or more. No one knows for sure. There were too many fires and many of them burned into one another. For the western settlers trying to escape from these fires, an official fire count didn't matter; they confronted a firestorm burning everything in its path.

It is remarkable more people did not die in the fires of 1910. The heroic stories of those who survived paint a horrific scene of devastation, as the fires burned through the towns of Taft, DeBorgia, Henderson, and Haugan in Montana, and the town of Wallace in Idaho. Firefighters avoid entering caves for safety because fire can draw out the relatively cooler air inside and replace it with smoke and hot air. However, in 1910, United States Forest Service Forest Ranger Ed Pulaski had no choice. To escape the Great Idaho Fire, he led his crew of 45 firefighters into a small abandoned mineshaft to escape the fire.[18] Pulaski stood guard at the entrance to beat back the flames that were burning the supporting timbers at the mine entrance and to prevent any of the frightened firefighters from running away.

A year after the 1910 Great Idaho Fire, Canada experienced a similar catastrophe when the Great Porcupine Fire ravaged northeastern Ontario. Cochrane, South Porcupine, and Pottsville were isolated but booming mine communities along the Canadian National Railway line in northern Ontario. Blinded by the rewards of the gold rush of 1908 and 1909, the settlers in these wood-built communities were defenseless from wildfire, their hands too busy exchanging money to get dirty from clearing the encroaching flammable forest. Nobody saw the scene of drought, hundreds of small land-clearing fires, carelessness, and very strong winds. Nor did they realize another great fire was about to take the stage.

On or about Wednesday, July 5, 1911, a fire started by unsafe lumber operations southwest of Cochrane grew in size and coalesced with several land-clearing fires. Settlers about 13 km west of Cochrane noticed the developing firestorm around 1000h on Tuesday, July 11.[19] Heavy clouds of yellowish smoke appeared first, followed by dense, black rolling clouds mottled red with burning gas. Panicked settlers fled their homes and rode into Cochrane on a narrow dirt road that followed the railway tracks. They raised the alarm as they rode west of Cochrane through tent cities, which housed an estimated 1,000 prospectors and their families.[20] Despite their warnings and the choking smoke, the smell of gold remained stronger than the smell of smoke. When the fire alarm finally sounded, it was too late. The firestorm hit hard and fast. Just before the 30 m wall of flames devoured the town, tornado-force winds laden with hot dust and debris engulfed the community. Darkness, fear, and chaos shrouded the town.

Within one hour, the communities of Cochrane, South Porcupine, and Pottsville were obliterated. Every building was destroyed except the Cochrane railway station. Using water from the adjacent water tower, the telegraph operator soaked the building and its cedar shingles. His actions saved the building and the hundreds of residents seeking shelter inside. The only other survivors were the people who fled into the lake and did not drown.

The 1911 Great Porcupine Fire also destroyed the communities of Porous Junction, Goldlands, and part of the town of Timmins. Thousands of people lost their homes. Although hundreds reportedly died, the final death toll is not known.

Disastrous fires struck again in northeastern Ontario in 1916 and 1922. On July 29, 1916, man and nature collaborated to cause the deadly 200,000 ha Matheson Fire. Numerous settler and lightning-caused fires merged into a deadly wind-driven conflagration that killed 223 people.[21]

Officially, the 1922 fire season had ended when, on October 4 and 5, the 168,000 ha Haileybury Fire destroyed the Ontario communities of North Cobalt, Charlton, Thornlee, and Heaslip. Without the need to obtain burning permits, settlers continued burning to clear land. They quickly lost control of the small fires to the very strong winds. Forty-three people lost their lives.[22]

Eight years after the 1910 Great Idaho Fire, one of the worst disasters in the history of Minnesota occurred on October 12. Sparks from a train engine ignited a massive fire in northeastern Minnesota. The statistics for the 1918 Minnesota Fire are staggering. The 101,170 ha fire destroyed 38 communities and 52,000 homes. A reported 453 lives were lost. Property damage estimates range from $7 to $30 million.[23] Surface winds carried the smoke from this fire eastward as far as Texas and South Carolina.[24] People observed and reported red and greenish-yellow suns, but there were no observations of a blue sun.[25]

George C. Joy with the Washington Forest Fire Association delivered an address in Portland, Oregon on November 25, 1922 at the Annual Conference of the Western Forestry and Conservation Association. The 1922 fire season was not just another bad fire season; losses to logging operations were unprecedented. Joy's address, titled "The Conflagration Hazard: Western Possibilities of Sweeping Fires Like Those of Minnesota and Canada," was published by the Oregon State Board of Forestry.[26] Concerned about the large area burned and the losses to the logging operators, the board pressed the government to take more responsibility as described in the board's forward to Joy's address:

> We are not yet on top of the fire problem. Once well started, with weather conditions favorable to spread of fire, control is always difficult; sometimes impossible. A review of what has happened in the past should sound a warning for the future. Adequately to safeguard our mature forests and allow our cut-over lands to reforest, much greater prevention effort must be exerted. While every citizen has an individual responsibility in this matter it remains for nation and state through example as well as through adequate laws and their enforcement to effect such a system of fire prevention as will guarantee against conflagrations with resulting loss of life and property.[27]

Joy made a passionate plea to increase fire prevention by increasing enforcement and the number of fire patrols. The previous 1902, 1910, and 1912 fire seasons resulted in significantly higher losses of life and timber, but the firefighting costs in 1922 were a record high. Joy's talk included a review of the most destructive wildfires in the United States and Canada. The list was long, too long, Joy felt, for it to continue. The 1825 Miramichi Fire topped Joy's list of destructive fires. Other fires followed, mostly in the west. The 182,000 ha Yaquina Fire in Oregon in 1846 burned 25 billion board feet of timber.[28] In May 1863, Pontiac Fire in Quebec burned about 648,000 ha.[29] In September 1868, smoke from the 121,000 ha Coos Fire in Oregon burned as far north as Portland.[30] The 1871 Peshtigo Fire in Wisconsin started in the evening of October 8. Although not the largest fire in the United States, it is the most disastrous; over 600,000 ha burned, and between 1,200 and 2,400 people died.[31] On October 5, 1881, a strong southwest wind in Michigan fanned numerous land-clearing fires into a large inferno known as the Great Fire of 1881. It burned over 400,000 ha and claimed 138 lives.[32] The September 1, 1894 Hinckley Fire in Minnesota burned only 64,750 ha but killed a reported 418 people and became Minnesota's second deadliest fire.[33] The death toll, however, was likely much higher. Officials did not know who was working, travelling, and living in the forest. The Hinckley Fire, like many of historic firestorms, started in the fall.

The 1902 Yacolt Fire in Washington claimed 38 lives. There were more than 80 fires that burned about 283,000 ha through August and September. Pushed by a strong southeast wind on September 11, the Yacolt Fire burned 96,685 ha and became Washington's largest fire.[34] The record still stands. Smoke from the Yacolt Fire and other fires burning in the region darkened the skies over southwestern Washington. At 1100h navigation for a steamer on the Columbia River could only be made using the boat's searchlights. Six days after the Yacolt Fire started, Ridgefield, Washington reported total darkness at 1500h.[35]

Joy finished his list of conflagrations by talking about the August 1, 1908 forest fire that destroyed the town of Fernie, British Columbia and killed 100 people.[36] He also described in detail the destruction caused by the 1910 Great Idaho Fire and the 1918 Minnesota fire.

There is little documentation on the massive prairie fires in Saskatchewan and Alberta during the hot and dry fire season in 1909. Roder reported an estimated area between 12 million hectares and 18 million hectares burned, but he makes no reference to any subsequently darkened skies.[37]

On May 21, 1919, the front page of the daily Edmonton newspaper the *Morning Bulletin* told the incredible story of another large firestorm in western Canada: "Lac La Biche Village in Ashes; Entire District is Homeless; Condition of People Perilous."[38] Except for the railway station, church and priest's house, and the McArthur's hotel, the entire village was destroyed on the Monday afternoon of May 19. Amazingly, all 300 people from an estimated 50 families survived the Great Fire of 1919. Many people took refuge in the same lake that kept the town alive by moving logs from the nearby McArthur logging camp to town and keeping the fish cannery busy. A delegation from Lac La Biche led by Father Okhuyson arrived in Edmonton on Tuesday afternoon seeking assistance. The *Morning Bulletin* news story reported what the delegation saw:

Although the fire which wiped out the town of Lac La Biche Monday came in the middle of the afternoon, it was black as midnight and the only illumination was from the fire itself, according to Father Okhuyson, who heads the delegation, sent down to Edmonton by the unhappy settlers to secure aid. The wind was blowing a terrible gale, states the Father, trees were bent level with the ground with its force and the air was as hot as to be insufferable for miles back. Teams coming into the village that afternoon were forced by the heat to turn back, when six miles out of town, and it was only possible to save the people in the town by covering them with wet blankets at the shore of the lake. Some idea of the intense heat may be gathered from the fact that the reeds on the shore of the lake were burned to the waters edge.[39]

The 1919–20 annual report for the Department of the Interior briefly mentioned a large fire in Saskatchewan, which burned an estimated 1.1 million hectares.[40] Although the report makes no mention of the area burned in Alberta, Peter Murphy estimates this may have been one million hectares.[41]

Robert Henry, director of the Dominion Forestry Branch, described the 1919 fire season as "the most disastrous ever for forest fires since the establishment of the Forestry Branch."[42] Theresa Desjarlais was a ten-year-old Metis child when her father stormed into their tent and screamed, "Fire!" She remembered that horrifying afternoon. "It was pitch dark but there was a yellow glow which seemed to reach the sky."[43] Fred Moses, a member of the provincial police force stationed in Lac La Biche, wrote in his diary for May 19, 1919, "cold, windy, eclipse of the sun, and Lac La Biche wiped out by flames."[44]

Eighteen dark days have been recorded in North America from 1706 to 1910.[45] High altitude smoke from forest fires or ash from volcanic eruptions caused most of these events (see Table 3.1).

Massive wildfires ravaged not just North America but countries throughout the world. The only requirement was, and still is, a source of fuel. Wildfire does not discern the nationality of the fuel. If it is combustible, it will fuel the wildfire. In 1915, after experiencing the worst drought since weather recording began in 1870, western Siberia burned. Within about 50 days, forest fires burned out of control in four million hectares of the largest boreal forest region in the world.[46]

TABLE 3.1 Dark days in North America, 1706–1910

Day	Duration	Area Affected by Darkness
May 12, 1706	1000h–unknown	New England
October 21, 1716	1100h–1130h	New England
August 9, 1732	Unknown	New England
October 19, 1762	Unknown	Southern Ontario and Detroit, MI
May 19, 1780	4 hours of darkness	New England ("Black Friday")
October 16, 1785	Unknown	Canada
July 3, 1814	Unknown	New England to Newfoundland
November 6–10, 1819	Unknown	New England and Canada
July 8, 1836	Unknown	New England
October 16, 1863	Brief duration	Canada
September 15–October 20, 1868	Unknown	Western Oregon and Washington
September 6, 1881	Unknown	New England ("Yellow Day")
November 19, 1887	Unknown	Ohio River Valley ("Smoky Day")
September 2, 1894	Unknown	New England
September 12, 1902	Unknown	Western Washington ("Dark Day")
June 5, 1903	Unknown	Saratoga, NY
December 2, 1904	1000h–1015h	Memphis, TN
August 20–25, 1910	Unknown	Northern United States from Idaho and northern Utah east to the St. Lawrence River

Large high-intensity fires burned a contiguous area of about 1.3 million hectares.[47] Less extensive burning (from smaller and likely less intense fires) occurred in the remaining area. Thick smoke blanketed 1.8 million hectares.[48] Unlike the 1950 firestorm in western Canada, the smoke from the 1915 Siberian fires remained concentrated regionally, resulting in reduced temperatures of 5 °C to 15 °C, and a 10- to 15-day delay in the harvesting of crops.[49] Compared to the 1950 firestorm, the 1915 Siberia fires burned over a larger area and dispersed more animals. Large migrations of mammals, including squirrels, bears, and moose occurred.[50]

The largest documented single fire in the world is the Great Black Dragon Fire. From early May to early June in 1987, this fire consumed virgin coniferous forests along the Amur River, bordering eastern Siberia and Chinese Manchuria. It burned over seven million hectares—over one million hectares in China and about six million hectares in Siberia.[51] This is the largest documented single fire in the world.

The circumboreal forests are truly home to the largest wildfires on earth. The energy that can be released from large wildfires was described by George Joy during his talk at the 1922 Western Forestry and Conservation annual meeting in Portland, Oregon: "Nine years later [after the 1902 Yacolt Fire] I went through an experience with a great crown fire, and I must add that until you have seen one—seen the huge, red streamers fling themselves up into the sky and the dense rolling clouds of black smoke rush upward toward the heavens—you can not fully appreciate how powerless is the hand of man to stay the forces of nature when she asserts herself with all her might."[52]

The only other documented widespread occurrence of a blue moon and blue sun sighting over a large area is from the 1883 eruption of the Krakatoa volcano.[53] This explosion was one of the worst natural disasters to occur on earth. The 45 km² island volcano of Krakatoa lies in the Sunda Strait between the Indonesian islands of Java and Sumatra. On August 27 at 0007h, the first major explosion occurred. Four additional explosions occurred the next morning.[54] After the last explosion at 1702h, the northern two-thirds of the island disappeared. The explosions and subsequent collapse of the volcano created giant tidal waves, some over 36.6 m (120 feet) in height. These tidal waves or tsunamis destroyed 165 coastal villages in the Sunda Strait and killed 36,417 people.[55]

The energy released by the eruption of Krakatoa was equivalent to 20,000 times the energy released by the Hiroshima atomic bomb.[56] The explosions were so powerful they were heard 3,540 km away in Australia. The amount of energy released easily thrust a large cloud of dust and ash into the stratosphere, 27 km above the island. For two days, darkness covered areas 80 km away. Once in the stratosphere, the dust and ash formed a band that circled the earth. By September 9, 1883, this persistent band of fine dust and ash particles completed its first orbit of the earth, thereby covering the entire upper atmosphere parallel to the equator. As the dust and ash continued circling the earth, it spread to higher latitudes. By December, many unusual atmospheric effects were observed, including the occurrence of red, green, and blue suns. These coloured suns were observed periodically for about three years.[57]

The 1883 Krakatoa eruption caused global climate changes by reducing the amount of solar radiation reaching the earth. The average global temperature dropped by 1.2 °C and did not recover to normal levels until 1888.[58] Extreme weather events (severe winter, heavy rains and floods, and drought) were also documented around the world.

The cooling of the climate that follows major volcanic eruptions such as the Krakatoa eruption and the 1991 Mount Pinatubo eruption in the Philippines has caught the attention of scientists interested in large-scale geoengineering to manage global warming. On June 15, 1991, Mount Pinatubo blasted 15 to 20 million tons of sulphur dioxide gas into the stratosphere. Sulphur dioxide reacts with water vapour to form small droplets of sulphuric acid. The Mount Pinatubo eruption created a dense layer of sulphuric acid droplets that blocked some of the sun's incoming radiation, and reduced global temperatures in 1992 and 1993.[59]

The smoke from the 1950 forest fires in northeast British Columbia and northwest Alberta also reduced temperatures. Forecasters in Washington, DC did not expect a 46% reduction in the amount of solar radiation reaching the ground, but from September 24 to 27, the observed maximum temperatures were about 3.5 °C lower than what was forecasted.[60]

Harry Wexler, chief of the Special Scientific Services Division at the United States Weather Bureau in Washington, DC, described the September 1950 smoke event as the "Great Smoke Pall" because the smoke covered such a large area in North America and was observed over the Atlantic Ocean and in northwestern Europe.[61] The high concentration of smoke reduced the incoming solar radiation and produced unique optical effects: darkness, coloured skies, moons and suns of varying colour and intensity, blue rays of light that flooded through windows, and unexpected blockage of the eclipse of the moon.

Contemporary fire management policies were, in part, shaped by the dark days of the past. After the 1910 Great Idaho Fire, the United States government decided forest fires should be fought. Congress subsequently loosened its purse strings and provided additional resources to the Forest Service. The big fires and big smoke during the 1950 fire season in western Canada became the catalyst for similar forest protection policy changes in Alberta. I discuss the lessons learned from the

1950 fires in the Conclusion, but next we turn to what is perhaps the most fascinating optical effect of the Chinchaga Firestorm and this volume's namesake: the blue sun and the blue moon.

FOUR

BLUE MOON, BLUE SUN

The first observation in Europe of a blue sun was reported from Bwlchgwyn, near Wrexham, Denbighshire in Scotland, just after the sun rose on the morning of September 26, 1950.[1] This observation was brief as clouds soon obscured the coloured sun. However, during the early afternoon, the sky began to clear and the blue sun reappeared. By late afternoon, the blue sun could be seen from Scotland, Wales, and western England. When the moon later appeared, it too glowed blue. One observer from Glasgow said the sun appeared as though he was looking at it through the deep blue, anti-glare glass installed in some automobiles. The magical lure of the blue sun even drew the attention of those indoors, as rays of blue light filled people's rooms.[2]

Because the smoke layer reduced the intensity of the sun, people watching the blue sun did not need to protect their eyes. Some people who called the weather office in Edinburgh, Scotland, did not know if what they saw was a moon or a sun.[3] Some may have wondered whether it was a new blue planet that had lost its way.

The Wednesday, September 27, 1950 edition of the *Times* newspaper in London, England, reported the occurrence of the blue sun. The *Times* weather correspondent who wrote the article suggested the reported high elevation layer of haze could have been produced by extensive forest fires or a volcanic eruption. The article included a correct explanation of the physical process that caused the colouration of the sun. The weather correspondent knew the number and size of the particles between the sun and the observer could change the colour of the sun as seen by the observer. The article described the requirement for small particles to create the often seen rising or setting sun, but for the rarely seen blue sun, a critical particle size was required.[4]

Alan Watson was flying north of Edinburgh on September 26, 1950 when he first noticed the blue sun. When Watson opened the cockpit canopy and removed his sunglasses, he knew something strange was happening. He radioed air traffic control at the Royal Air Force (RAF) Base in Leuchars to report the blue sun.[5] A RAF Meteor aircraft was immediately dispatched to a height of 13,106 m (43,000 feet) to investigate this phenomenon and report on any colour changes during the aircraft's ascent. At 9,449 m (31,000 feet), pilot West Jones reported entering a cloud of brown haze at 1549h. The sun, he noted, still appeared blue.[6] Another pilot flying near Cambridge reported a strong smell of smoke in the cockpit of his aircraft as he flew through the same layer of brown haze between 9,144 m (30,000) and 9,754 m (32,000 feet).[7]

Jones climbed another 2,134 m (7,000 feet) and reported that he was almost through the layer of haze obscuring the ground below. The sun appeared a normal colour but with a tinge of blue. At 13,106 m (43,000 feet), Jones reported flying above the layer of haze. At this altitude, the sun shone bright yellow, without any hint of the phenomena occurring below.[8]

Before Jones flew the surveillance plane back to report his observations, the rumours had already spread quickly. According to Gerrard

Faye, a reporter from the *Manchester Guardian*, "Good people all over the north were asking each other if this was not either some retribution descending on the universe for our atomic meddling, or even worse, the first sign that the earth was about to do something eccentric, like freezing solid or blowing up."[9]

Faye reported three theories. The first theory suggested dust or ice particles interfered with the red portion of the visible spectrum of light, thereby allowing the blue light to pass through. This theory correctly explained the physics, but it did not identify the correct particle type. The second theory associated a layer of haze of unknown origin, at about 9,144 m (30,000 feet), with filtering the light from the sun and moon. This theory was confirmed by pilots who encountered a layer of haze between 9,144 m (30,000 feet) and 12,192 m (40,000 feet). Smoke from large forest fires in British Columbia was identified as the third theory.[10] Although this theory was also correct, the smoke also came from forest fires burning in Alberta. All three theories provided important clues to solve the blue moon, blue sun puzzle.

Faye ended his report by stating he did not support any of these theories. Rather, he concluded, "I believe we're entitled to a blue moon now and then, at least, once in a blue moon."[11]

The Air Ministry Meteorological Office in Scotland concluded, "It was thought the phenomenon was caused by the diffraction of light by dust particles of a certain size in a haze layer between six and eight miles high. It is possible that these dust particles originated from forest fires in North America, and were carried across the Atlantic in strong upper air currents."[12] When Jones returned to the RAF Base he brought back evidence to confirm the Air Ministry's speculation of the cause of the blue sun and moon. An oily film covered the aircraft and inside the cockpit the smell of smoke still lingered.

In Arbroath, Scotland, just after 1600h, the sun started to gradually change colour from yellow to gold and then from silver grey to indigo blue by 1630h. Pedestrians in town stopped to stare at this unusual spectacle. For several hours after the blue sun disappeared, inquiries to the newspaper offices jammed the phone lines.[13]

When Dr. Mary Jackson returned to England for a visit in January 1952, many people spoke to her about the appearance of the blue moon in 1950. They wanted to know more about the monstrous forest fire that produced so much smoke and how the small settlement of Keg River survived the fire.

A bluish sun and moon were also observed in France, Belgium, Portugal, Denmark, and Switzerland on September 26.[14] People in Netherlands and Norway saw the blue sun as it rose the next day.[15] In Gibraltar, blue sun observations lasted until the morning of September 30.[16]

Special conditions were required for the smoke to be able to obscure the sun and moon for such a long time and over such a large area. These same conditions also created the many different optical effects. Although numerous blue moon and blue sun sightings were made, the reported colours ranged from dull red, copper, orange, green-yellow, pink, silver, purple, blue violet, and lavender. Most of the sightings, however, were of purple, blue violet, and lavender. At Grand Rapids, Michigan, the sun appeared blue with a yellow aura.[17] In Philadelphia, a beautiful lavender sun appeared. Observers there also reported the moon as "purple at times, otherwise of normal colour."[18] In Baltimore, the moon was reported as "bright white, sometimes dim, with an occasional purple cast."[19] The moon had a "pronounced bluish tinge" at Newark, New Jersey, and at Concord, New Hampshire, observers reported an unusual grayish cast moon.[20]

The phrase "blue moon" first appeared in literature, in a poem published anonymously in 1528, titled "Rede Me and Be Nott Wrothe."[21] Lines 3,346 and 3,347 of this early English writing read,

> Yf they say the mone is blewe
> We must believe that it is true.

Jerome Barlowe and William Roye, two expatriate English friars, are credited with writing this indictment of the Roman Catholic doctrine and practices and the church's leader, Cardinal Thomas Wolsey.[22] They mocked the church leaders by suggesting the common folk should blindly follow and believe them if they say the moon is blue. A blue moon in this context refers to an absurdity or an improbability.

Over time, this meaning changed to describe an event happening "very infrequently" and then to "once in a while."[23] The current but misinterpreted second moon theory describes a blue moon as the second full moon to occur in a calendar month.[24] This phenomenon occurs, on average, once every 2.72 years. Only 2.977% of all full moons are also blue moons.

The time between successive full moons is called a lunar or synodic month. The 29.531-day lunar month is not synchronized with the Gregorian calendar. February has 28 days. April, June, September, and November have 30 days, and the remaining months have 31 days. This equates to 365 days. Since it takes about 365 and a quarter days for the earth to orbit the sun, an extra day is added to the month of February for every year divisible by four. This is called a leap year. The Gregorian calendar, however, is not perfect. The time it takes the earth to make one revolution around the sun is just less than 365 and a quarter days. Adjustments are made to account for this difference, which, although small, accumulate over time. A year divisible by 100 is not considered a leap year, unless it is again divisible by 400. The year 2000 was, therefore, a special leap year.

Folklorist Philip Hiscock, an archivist at the Folklore and Language Archive at Memorial University in St. John's, Newfoundland, traced the origins of the second full moon theory to the Canadian-designed board game Trivial Pursuit.[25] The 1986 Genius II edition of Trivial Pursuit included the question "What is a Blue Moon?" The answer—a second full moon within a calendar month—was obtained from the 1985 publication the *Kids' World Almanac of Records and Facts* compiled by Margo McLoone-Basta, Richard Rosenblum, and Alice Siegel.[26] These compilers, in turn, got the second moon theory from a late January 1980 National Public Radio program called *Star Date*, hosted by Deborah Byrd.[27] *Star Date* got the second moon theory from an article on blue moons written by James Pruett in the March 1946 publication *Sky & Telescope*.[28] Pruett, an amateur astronomer from Eugene, Oregon wrote, "Seven times in nineteen years there were—and still are—13 full moons in a year. This gives 11 months with one full moon each and one with two. This second in a month, so I interpret it, was called 'Blue Moon.'"[29]

Pruett repeated this misinterpretation from a question and answer column written by Laurence Lafleur in the July 1943 issue of *Sky & Telescope*.[30] Lafleur called the 13th full moon in a year a blue moon.

According to Lafleur, the second moon theory was taken from the 1937 *Maine Farmers' Almanac*. From this point, two separate definitions evolved for the meaning of blue moon. The story is described in another *Sky & Telescope* article written in May 1999.[31] The authors, Roger Sinnott, associate editor of *Sky & Telescope*, Donald Olson, a professor of physics at Southwest Texas State University, and Richard Fienberg, publisher of *Sky & Telescope*, searched through more than 40 editions of the *Maine Farmers' Almanac* from 1819 to 1962. They found more than a dozen references to blue moons. Interestingly, not one of them was a second full moon in a month. They knew this because the blue moons in the *Maine Farmers' Almanac* occurred on the 20th, 21st, 22nd, or 23rd day during the months of November, May, February, or August. Since a lunar month is 29.531 days, it is not possible to have two full moons in the same month, if the first full moon occurs on the 20th, 21st, 22nd, or 23rd day. What, then, did the blue moons refer to?

Olson and his colleagues found a link between the blue moons in the *Maine Farmers' Almanac* and the seasons. In the northern hemisphere, the summer solstice is the longest day of the year (approximately June 21), and the winter solstice is the shortest day of the year (approximately December 21). Spring or vernal equinox occurs when the night and day lengths are the same (approximately March 21). Fall or autumnal equinox occurs on the first day of autumn when the sun crosses the celestial equator (approximately September 22). The blue moon dates in the *Maine Farmers' Almanac* occur about one month before the corresponding dates of the solstices and equinoxes.

Instead of using the familiar 365-day calendar year, the *Maine Farmers' Almanac* uses the solar calendar based on the tropical year, defined as the time between the repetitive occurrences of the winter solstice. A tropical year is slightly shorter than a calendar year. Each season within a tropical year typically contains three full moons, and each full moon is named based on when it occurs relative to the solstices and equinoxes. Occasionally, a tropical year contains an extra full moon. This means that one of the seasons experiences four full moons. When this occurs, the third full moon during this season is called a blue moon. The reason for this is simple. All of the other full moons have pre-assigned names. To maintain the naming synchronization when four full moons occur in a season, the third full moon becomes a marker full moon and is called the blue moon. The blue moons were distinguished from the other full moons on the calendar by being a solid colour.

There are also two different definitions of the start of the seasons, which yield different dates for a traditional blue moon. Astronomers use the actual position of the sun to define the start of each season. Since the earth's orbit is not circular, the astronomical seasons are not equal in length. In the northern hemisphere, the spring and summer seasons are longer than the autumn and winter seasons. The *Maine Farmers' Almanac* definition of a blue moon uses a different method to calculate the start of the season. This method, called the Right Ascension of the Mean Sun, is based on a fictitious sun travelling at a constant speed. This results in seasons of equal length and blue moons that occur on fixed dates.

The 1943 issue of *Sky & Telescope* planted the seed for the misinterpreted meaning of the term *blue moon*. The *Star Date* program and the Trivial Pursuit board game allowed the second moon theory to grow and become established in our vocabulary. The phenomenal world media coverage of the occurrence of a second full moon in May 1988, and second full moons in January and March 1999, provided full tenure for the phrase *blue moon*.

Three years after the Chinchaga Firestorm, Rudolf Penndorf from the Geophysics Research Directorate at the Air Force Cambridge Research Center in Cambridge, Massachusetts, wrote a technical report to explain the phenomenon of the blue moon and sun.[32] Penndorf concluded that a consistent and unique particle size at very high concentrations was required to selectively scatter more of the longer wavelengths of visible light (green and red) and less of the shorter wavelengths of visible light (blue and violet). Using Mie's theory of selective large particle scattering, he determined that the size of the smoke particles over eastern United States and Europe must have been between 1.0 μm (micrometre) and 1.6 μm in diameter. A micrometre (also called micron) is one-millionth of a metre, or about one-twentieth the diameter of a strand of hair. Penndorf also estimated the density in the smoke layer to be about 127 to 175 particles per cubic centimetre.

Light scattering in all directions from air molecules (oxygen and nitrogen) and very tiny particles (diameter less than one-tenth that of the wavelength of light) is referred to as Rayleigh or Molecular Scattering.[33] Not all wavelengths of light scatter equally. Rayleigh's law states that the scattering coefficient for air molecules is inversely proportional to the fourth power of the wavelength ($1/\lambda^4$). Since the colours violet and blue in the visible portion of the electromagnetic spectrum are the smallest wavelengths, they scatter the strongest across the sky. For example, violet light at $\lambda = 0.40$ μm scatters 9.4 times more than red light at $\lambda = 0.70$ μm, hence, the blue colour of the sky. Generally, bluer skies indicate fewer particulates and cleaner air. In comparison, large water droplets (20 μm in diameter) in clouds scatter all visible wavelengths. Clouds appear white because the equal mixture of all colours produces white light. The absence of light yields a black colour.

The spectacular sunrises and sunsets we sometimes see are due to both Rayleigh and Mie Scattering. The lower angle of the sun at sunrise and sunset increases the light's travel path, which intensifies the effects of scattering, including multiple scattering. The farther the particles of light travel through the atmosphere, the less likely they are to miss molecules. Air molecules and fine particles therefore scatter most of the blue colour out of the ray of light. Since red light scatters less, it passes through the atmosphere and reaches our eyes. Sunsets are typically more intense than sunrises; the troposphere at sunset holds more particulates than at sunrise. When the sun is overhead, its rays of light travel their shortest path. Thus, at midday, all wavelengths reach the earth with nearly equal intensity, resulting in the sun appearing yellow-white.

Red is a good colour choice to signify danger. Red railway crossing lights and car taillights, for example, are easier to see from afar, at night, and in the fog, compared to the shorter wavelength colours. This is due, in part, because red light scatters the least.

Rayleigh Scattering explains why the sky is blue but not why the moon and sun appeared blue in 1950. Nor does it explain the vivid red sunrises and sunsets. The redirection of light by Rayleigh Scattering is caused not by the electromagnetic wave bouncing off particles but by its interaction with them at the molecular level in a process called induced dipole movement. Large particles resist this elastic scattering (incident light and redirected light have the same frequency) because they are too big to create a separation of electrical charge that is possible with small particles. When particle sizes in the atmosphere are nearly equal to or larger than the wavelength of visible light, a more complex process called Mie Scattering (also called Lorenz-Mie Scattering) takes over.

Large spherical particles of similar size selectively scatter light in a predominant forward direction. Although Rayleigh Scattering accounts for the brilliant yellow-orange colours of the sun at sunrise and sunset, Mie Scattering from larger particles such as smoke or dust enhances these colours. The larger particles scatter all wavelengths of light within the visible portion of the electromagnetic spectrum, but red and orange are scattered less efficiently than blue and green. This contributes to the occasional crimson red of the setting sun. The redness of the sun increases in intensity as the concentration of atmospheric particles increase. Mie Scattering explains why firefighters take some of the best sunset photographs. It also explains the rare observations of blue suns and moons.

The specific colour of the moon or sun depends on the size and density of the particles in the atmosphere. When the wavelength is half the size of the particle, Mie Scattering is at a minimum. Scattering reaches a maximum when the wavelength and particle size is the same, then diminishes to zero at longer wavelengths. Blue moons and blue suns occur when there are high concentrations of particles about 0.8 μm to 1.2 μm in diameter. A greenish moon or sun occurs when there are high concentrations of particles 1.8 μm to 2.0 μm in diameter.[34]

The preponderance of observations of purple, blue, violet, and lavender suns and moons in 1950 from September 24 to 27 suggests that the smoke particles were of a sufficient size to effectively scatter out more of the green and red portions of visible light in the electromagnetic spectrum. This allowed the shorter wavelength blue light to pass through.

Other factors influence the colours we see. In the visible portion of the electromagnetic spectrum, the longer wavelengths (red) are the lowest in energy. The sun's radiant energy peaks in the shorter wavelengths (blue) and decreases to its lowest intensity in the longer wavelengths (red). Despite the stronger energy associated with blue light, we are less capable of perceiving this colour because our eyes and brain bias the green and yellow wavelengths.

Absorption by particles is another factor. Particles can absorb and remove energy from a beam of light as it traverses through the atmosphere. The removal of energy by the absorption and scattering of light by particles is called extinction. Since absorption causes extinction, it can also change the colours we see. Grass is green because the chlorophyll in the leaves absorb all visible light except the wavelengths corresponding to the colour green. Green is reflected and perceived by the photoreceptor cells in the retina of our eyes. The rich blue colouring of the sky around sunset is also a result of absorption. The ozone layer selectively absorbs orange light and allows blue light to pass by.[35]

The sightings of blue suns in 1950 helped launch Sir Robert Wilson's career as a renowned British scientist. He was intrigued by the occurrence of the blue sun throughout Edinburgh from 1500h until sunset on September 26. While his colleagues at the Royal Observatory frantically answered phone inquiries from the public, Wilson quietly disappeared to use the observatory telescope to obtain a spectrogram of the blue sun. A spectrogram provides a characterization of the electromagnetic radiation the earth receives from the sun. He published his findings in 1951 in the *Monthly Notices of the Royal Astronomical Society*.[36] The spectrogram showed a high extinction of the longer wavelengths (red light) of the electromagnetic spectrum. The scattering of electromagnetic radiation by the smoke particles caused this extinction. For the type of scattering measured by Wilson to occur, a uniform size distribution of particles is required.

Wilson later applied the scattering theory he used in his blue sun work to study interstellar polarization and extinction. He also completed a critique of the effects of earth-orbiting "needles" on optical astronomy. In 1961, the United States Air Force dispersed 480 million copper needles, 1.78 cm long and 0.0018 cm in diameter, into space, to orbit as a ring around the earth at a height of about 3,500 km. The circular band of the tiny copper needles were intended to behave as an antenna to reflect radio signals and provide communication support for troops overseas. The needles failed to disperse. Success was attained two years later when 350 million similar, but thinner, needles were dispersed. Robert Wilson and other scientists around the world were concerned about the global impact of the needles, and in particular, their optical interference. In the course of this work, Wilson discovered two new diffuse interstellar bands.

Particulates less than 2.5 μm in diameter account for over 80% to 90% of all particulates from wildland fires.[37] Less than 10% of the particulates are between 2 μm and 10 μm in size.[38] The initial size distribution of the particulates depends on the intensity of the fire and the atmospheric conditions, but particles less than 1 μm in diameter usually dominate. Approximately 99% of the particulates measured in smoke plumes from three experimental crown fires in mature jack pine stands were less than 1.2 μm in diameter. The median particle size diameter was 0.47 μm.[39]

In the 1980s, a heightened interest in the global effects of nuclear war lead to the initiation of new research projects to examine the factors contributing to changes in particle size over time and how far particles travel. When particle concentrations and size distributions change, the optical effects also change. William Porch at the Lawrence Livermore National Laboratory in Livermore, California was interested in better understanding the importance of the distribution and concentration of submicron (<1 μm in diameter) and supermicron (>1 μm in diameter) particles, and the intensity and height of turbulence in the active smoke plume of a fire. Porch and his research team developed a plume aerosol coagulation model to simulate the atmospheric optical effects of aerosols generated from large wildfires.[40] Simulation models help explain why certain phenomenon occur. The inputs to these models can be changed to assess their relative importance.

Porch and his colleagues used their model to provide another important piece of the blue moon, blue sun puzzle. It is a process called coagulation. Turbulent processes in the active smoke plume cause small particles to be scavenged by larger particles. Their modelled data, combined with observations from prescribed fires, suggest submicron scavenging occurred during the 1950 firestorm in western Canada. Since small particles travel at relatively higher velocities, they collide and combine with slower moving larger particles. The scavenging of very small size particles (0.2 μm in diameter) and the sedimentation of the larger particles resulted in an increase of the number of particles around 1 μm to 2 μm in diameter in the smoke plume. When particles attain a size of around 1 μm to 10 μm in diameter, their collision efficiency and subsequent growth by coagulation drops despite continued turbulence.[41] This is called the Greenfield Gap.[42] Because particles greater than 2 μm to

3 μm in diameter are too heavy, they eventually sediment out of a smoke plume. This leaves particles between 1 μm and about 3 μm in diameter. Aerosols tracked from atmospheric nuclear explosions have the same persistent size distribution. The Greenfield Gap concerns scientists because these stubborn aerosols can have global impacts.

Porch concluded that the initial concentration of smoke particles less than 2 μm (submicron) in diameter and the size of the fire are the most important conditions to cause blue moons and blue suns.[43] While these conditions are critical, they alone do not complete the puzzle because large forest fires in the past did not result in sightings of blue moons and blue suns. Why was the 1950 firestorm in western Canada different?

The researchers at the Lawrence Livermore National Laboratory published their work 34 years after Penndorf provided the first critical piece of the puzzle: the requirement for high concentration of particles around 1 μm in diameter. This density requirement is important, as without it there is no selective scattering to overcome molecular scattering. Western Canadians did not see a blue moon or blue sun. It took several days of transport to change the size distribution of the smoke particles and subsequently alter the light scattering. Aerosols in the atmosphere nucleate at very small sizes (< 0.2 μm in diameter). During transport, particle settlement, coagulation, and the formation of cloud droplets helped to create the perfect particle size distribution when the smoke reached Washington, DC, after three days of travel. Porch also suggested the near saturation relative humidity contributed to particle growth.[44]

The submicron concentration is more important than the supermicron concentration because the larger particles scatter all visible light, and hence do not contribute to any selective scattering. The larger particles also settle out of the atmosphere faster. Small particles, in comparison, have a higher surface area to volume ratio and stay airborne much longer. Over time, the very small particles decrease in number because of the process of coagulation and cloud processing. Strong upper winds and the lower and upper inversion layers entrained the smoke in a distinct layer. This minimized dilution from vertical and horizontal dispersions and increased particle collision and subsequent coagulation. The large volume of smoke produced by all of the forest fires over several days also ensured the smoke sandwich was well fed.

It is unusual to include volcanoes and nuclear explosions in a discussion about forest fires, but the 1950 Chinchaga Firestorm in western Canada was like no other. Over 100 fires produced a tremendous volume of smoke—enough to selectively scatter light and dominate the optical effects from molecular scattering. The very high initial concentration of smoke was a critical requirement but, without the later high density of particles of uniform size, there would be no blue moons or blue suns. The occurrence of uniform-sized particles in the atmosphere is a rare event. This was the perfect firestorm—high smoke production, strong winds, double inversion, and the processing of the exact number and size of particles to cause a bluing of the moon and sun. The long travel distance of the smoke, as an aerial river from its source, allowed for the continuing precipitation of the larger particles and the subsequent retention of a uniform critical particle size.

Light scattering by particles can be demonstrated in simple experiments. To replicate a red sun, place a half-full glass beaker or glass bowl on an overhead projector. Turn the projector on. Dip a small, wet stir stick into a container of coffee creamer and then into the water. A change in the colour of the light projected through the water will occur. Continue to slowly add small amounts of creamer until the colour of the light no longer changes.[45]

Physics Professor Donald Huffman demonstrated a similar experiment to his astronomy students at the University of Arizona to explain why the light observed from stars is often reddened. Instead of water and coffee creamer, Huffman used an empty beaker and cigarette smoke. He blew cigarette smoke into a beaker placed upside down on an overhead projector. The projected light did turn red in colour as planned, but then it unexpectedly turned blue in colour. After some further experimenting, Huffman and colleague Craig Bohern made an interesting observation. The colour of the projected light depended on how long they held the cigarette smoke in their mouths.[46]

The longer the smoke particles are held in the mouth, the more time they have to collide, join, and grow in size. This process, described earlier in this chapter, is called coagulation. The mouth is a perfect environment for this experiment because the high level of water vapour supports smoke particle growth by coagulation.

Although site-specific and short in duration, a few blue moons have been observed. Every 30 seconds, a live image of the pyramids in Giza, Egypt is displayed on the website *PyramidCam.com*. A rare occurrence of a blue sun was captured on December 14, 2006.[47] Farmers burning stubble fields caused a smoke event similar to the Great Smoke Pall of September 1950 but on a much smaller scale. Smoke particles between the camera and the pyramids selectively filtered the red light and allowed the blue light to pass through. The uniform fuel type, travel distance, and atmospheric conditions likely contributed to an initial uniform particle size.

Deserts are an exception to the rare occurrence of uniform particle sizes in nature. Strong winds blowing across the Mongolian and Saharan deserts pick up tiny, round dust particles. Once carried aloft, the particles can travel long distances and produce green and blue suns. In Beijing, China, visiting scientists from the University of Arizona observed a blue sun in 1978 and again in Nanjing, China in 1979.[48] Their hosts casually mentioned the frequent occurrence of a blue sun when wind-blown dust from the deserts of Mongolia arrive. Although desert dust storms still occur, the observation of blue suns have faded, likely because of the increasing concentration of competing particles in the atmosphere from China's rapid industrial growth.

In wildland fire environments, the fuel complex (type, amount, size, and arrangement) can vary considerably across the landscape. The particles in the smoke column from wildland fires are therefore of varied sizes. For this reason, the Great Smoke Pall of 1950 and the widespread observations of blue suns and blue moons are significant. These phenomena are not known to have occurred before 1950 as a result of forest fires.[49]

FIVE

THE BIG SMOKE

Particles suspended in the atmosphere are called aerosols. They can be solid or liquid, and natural or human-made. Aerosols manufactured by humans account for approximately 10% of the total aerosols in the atmosphere.[1] Aerosols usually remain airborne for several hours to a day. Fine aerosols, particularly the non-water soluble aerosols, can have residence times of 20 to 80 days.[2] Rain scrubs the air and eventually removes these aerosols from the atmosphere. This is why the air is so fresh and clear after a rain shower.

   Five levels of magnitude from 0.01 μm to 100 μm in diameter are used to describe aerosols and their associated health risk. Aerosols less than 100 μm in diameter are classified as inhalable. They enter the respiratory system via the nose and mouth. A portion of these aerosols (less than 10 μm in diameter) pass the larynx and enter the airway (trachea) and the bronchial region of the lungs. The smaller aerosols (less than 4 μm in diameter) penetrate further into the smallest branches of the lungs called alveoli.[3] Alveoli are individual air sacs where the body's blood supply and the air we breathe exchange oxygen and carbon dioxide.

Wind-blown dust and salt from sea spray produce aerosol particles larger than 1 μm in diameter. The combustion of fossil fuels and biomass burning usually produce soot and smoke particles smaller than 1 μm in diameter. The release of sulphur dioxide (SO_2) during volcanic eruptions also produces particles smaller than 1μm in diameter. Sulphur dioxide reacts with water to form sulphate particles. After a 30-minute downpour on June 25, 1919, the Town of Dawson, Yukon was streaked with yellow from sulphur released during a volcanic eruption in Indonesia. The sulphate aerosols were transported in the stratosphere to Alaska and Yukon. When these aerosols re-entered the lowest layer of the atmosphere, called the troposphere, precipitation carried them back to earth. The sulphate aerosols from the 1919 volcanic eruption were detected in an ice core taken in 1996 from the Eclipse Icefield in the St. Elias Mountains in Yukon.[4]

Most aerosols remain in the troposphere, an 8 km to 18 km layer of the atmosphere that blankets the earth. The shortest heights of this layer occur at the North and South Poles where the air is colder and denser. Since the troposphere houses the earth's weather, the air in this zone is constantly mixing. Above the troposphere lies the stratosphere, a stable layer of air extending to a height of approximately 50 km. With increasing height, temperature decreases in the troposphere but increases in the stratosphere. In temperate latitudes, commercial aircraft pilots fly in the lower stratosphere at an altitude of about 10 km to avoid the weather. Because of the limited vertical mixing, aerosols finding their way into the stratosphere can circulate around the earth for many months, even several years.

When the volcano under Iceland's Eyjafjallajökull Glacier erupted on April 14, 2010, it violently spewed pulverized rock, glass, water vapour, and other gases to a height of 8 km to 10 km above sea level. The ash cloud grew quickly and covered a large area over Europe on

April 15. Scientists assumed these volcanic eruptions were the only natural force capable of producing enough energy to lift a smoke column and penetrate the lower stratosphere. Michael Fromm, a senior scientist at Computational Physics Inc. in Springfield, Virginia, and co-authors challenged this assumption in a letter published in *Geophysical Research Letters* in 2000.[5] They suggested a link between high-intensity boreal wildfires and the transport of aerosols into the stratosphere. During July and August 1998, smoke columns from boreal fires in Canada and Russia penetrated six kilometres into the stratosphere. By early fall, the smoke travelled around the globe. Satellites helped to make this remarkable new discovery. The United States Naval Research Laboratory's Polar Ozone and Aerosol Measurement (POAM III) instrument on board the SPOT-4 satellite measured atmospheric ozone, water vapour, nitrogen dioxide, aerosol extinction, and temperature. Because vertical profiles can be constructed from the POAM III measurements, smoke layers can be distinguished. NASA's Stratospheric Aerosol and Gas Experiment (SAGE II) instrument on board the Earth Radiation Budget Satellite also measured the distribution of aerosols, ozone, water vapour, and nitrogen dioxide. Ground-based Light Detection and Ranging (Lidar) measurements in Europe provided important corroborative evidence.

In 2003, Michael Fromm and René Servranckx, a meteorologist with the Meteorological Service of Canada, provided proof of the link between boreal fires and super cell convection in a published letter, again in *Geophysical Research Letters*.[6] They studied an extreme convective event on May 28, 2001 from the Chisholm Fire in Alberta. During the summer of 2001, POAM III made over 200 observations of aerosol layers in the stratosphere. Back trajectories of some of the layers traced the smoke to the 116,000 ha Chisholm Fire.

The search for super cell convections from boreal wildfires surprised the researchers. This phenomenon occurs more frequently than they first expected, which suggests that volcanoes are not the sole transporter of aerosols from the troposphere to the stratosphere.

In Kluane National Park in southwest Yukon, Mount Logan rises 5,959 m above sea level. It is Canada's highest natural sampling station to intercept passing aerosol plumes that contain particulates and

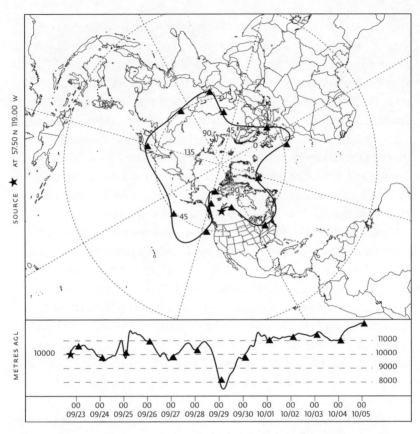

FIGURE 5.1 Trajectory of smoke from the 1950 Chinchaga Firestorm, September 22, 1950, 1100h based on a NCEP/NCAR Reanalysis. Distances in metres are shown as above ground level (AGL).

chemical compounds. During precipitation (dry snow or super-cooled water droplets that freeze immediately on contact with cold surfaces and form granular ice), chemical salts of sulphate, nitrate, sodium, and chloride, and dust, particulate matter, trace metals, and even diatoms are scavenged. As snow accumulates on Mount Logan, the deep layers transform into ice and preserve samples of past atmospheric conditions. The deposited aerosols remained locked in the snow and ice until Gerald Holdsworth, then a glaciologist from Environment Canada, arrived in 1980 and retrieved snow and ice cores on the upper plateau of Mount Logan at an elevation of 5,340 m.[7]

Volcanic activity produces high levels of sulphate (H_2SO_4 and partly neutralized SO_4); fossil fuel consumption produces high levels of sulphate as well as nitrate from nitrous oxide (N_2O). Lightning within thunderstorms also generates nitrogen oxides (NO_x) and hence nitrates in precipitation, but around Mount Logan lightning events are rare because of the extremely cold temperatures.

Holdsworth found a correlation between intense atmospheric thermonuclear weapons testing (ATWT) and spikes in radioactivity levels and nitrate concentrations measured in his glacier cores. The ATWT explosions generated large amounts of nitrogen oxides (NO and NO_2). After further oxidation, these gaseous nitrogen emissions form nitric acid when absorbed in atmospheric water vapour or adsorbed onto snowflakes. The cores collected by Holdsworth revealed a nitrate pulse in 1950. In 1945, the United States began atmospheric nuclear testing (one test in 1945, two in 1946, and three in 1948). The Soviet Union conducted one atmospheric nuclear test in 1949. The 1950 nitrate spike, which exceeded the 1963 spike, cannot be explained by volcanic activity or atmospheric nuclear testing. In 1950, no known volcanic eruptions or atmospheric nuclear tests occurred. What, then, caused the 1950 nitrate pulse in Holdsworth's ice cores?

Holdsworth recognized that the nitrate chemical signature could correspond to a forest fire event. Like ATWT explosions, wildfires produce significant nitrogen oxides emissions as nitric oxide (NO) and nitrogen dioxide (NO_2). These gases constitute the largest gaseous emissions into the atmosphere after carbon dioxide (CO_2) and carbon monoxide (CO). In the late 1980s, Holdsworth contacted Peter Murphy at the University of Alberta in Edmonton. Murphy felt it was possible the smoke event from the 1950 Chinchaga Firestorm caused the nitrate spike in his ice core. He contacted Stan Shewchuk, a meteorologist with the Saskatchewan Research Council in Saskatoon, and asked if he could do a forward trajectory of the smoke from the fire site to Yukon. Some of the trajectories, then completed by hand, came very close to Mount Logan.[8] These rough trajectories did not provide concise proof to link the nitrate spike with the Chinchaga Firestorm, but the evidence pointed very strongly in that direction.

While completing his PhD studies at the University of Toronto, Robert Field modelled the trajectories of the Chinchaga Firestorm smoke at different heights using a general circulation model. The trajectory using a 10 km injection height shows the smoke travelling around the northern hemisphere and finding its way back to Canada over Yukon (Figure 5.1). The trajectory modelling suggests the nitrate deposited on Mount Logan could have come from the September 1950 forest fires in western Canada.

Vegetation consumed in wildland fires typically contains low levels of nitrogen. Despite a low emission factor (1 kg biomass produces 0.002 kg or less of nitrogen oxides), wildfires can produce significant levels of nitrogen oxides emissions when large areas are burned.[9] The atmospheric lifetime of nitrogen oxides (NO_x) is about 10 to 15 days in summer (latitude of about 54° and an approximate altitude of 300 mb). During winter it is about 2 to 5 days.[10] Nitric acid typically deposits out within one day, longer if it is taken up on the surface of particles (aerosols).[11] It is therefore possible the nitric acid remained in the smoke layer for a longer than normal period before deposition.

Wildfires burning in Alaska could also have contributed to the strong nitrate spike in the ice core. The Schuman House–Porcupine River Fire, Little Black River Fire, and the Venetie Fire burned 695,493 ha in Alaska in 1950. These fires were detected on May 24 and reported as extinguished on September 22, 23, and 28, respectively.[12]

The district forester blamed muskrat trappers for causing the Schuman House–Porcupine River Fire and the Little Black River Fire. They often started smudge fires to repel mosquitoes, and marsh fires to burn the vegetation and force muskrats out of their homes. The Venetie Fire was the only one reported as a lightning-caused fire.

On September 14, 1950, smoke blanketed Edmonton. Alberta Forest Service officials reported the smoke came from 17 forest fires burning in northern Alberta and large forest fires burning in Alaska.[13] The deposition of nitric acid on Mount Logan could have occurred as a result of the Alaska fires in 1950; however, the required north or northwest wind flow is not a normal summer wind direction for that region. A preliminary investigation of daily vector wind plots using the National Centers for Environmental Prediction (NCEP)/National Center for Atmospheric Research (NCAR) Reanalysis model supports this claim.[14] It thus remains a curious possibility the 1950 fires in western

Canada contributed to the Mount Logan nitrate spike after circum-hemispheric transport.

Aerosols act as "seeds" or "nuclei" to start the process of condensation of small water droplets in the cloud. This is the rationale for cloud seeding. The formation of cloud droplets is very difficult without the presence of aerosol particles. Without aerosols, clouds would not exist.

The condensation or water vapour trails left by the exhaust from jet aircraft engines are called contrails. Wing tip vortices can also cause these artificial clouds. Water vapour is a powerful greenhouse gas. Clouds and water in the gaseous phase account for about 66% to 85% of the total natural greenhouse effect.[15] As the atmosphere warms, its ability to retain moisture increases, resulting in a positive feedback that further increases the temperature. Contrails only cover an estimated 0.1% of the earth's surface, but some regions, such as southern California and east central France, have much higher areas of coverage.[16] When air traffic stopped for three days following the September 11, 2001 (9/11) terrorist attacks on the World Trade Center in New York City, the contrails over the United States disappeared. This provided an unprecedented opportunity to study the effects of contrails on temperature. David Travis and Ryan Lauritsen from the University of Wisconsin-Whitewater and Andrew Carleton from Pennsylvania State University co-authored an interesting paper in *Nature* in 2002.[17] They compared the temperature range data (maximum temperature minus minimum temperature) from 4,000 weather stations across the United States for the period 1971 to 2000, with the temperature range data for the three-day contrail-free period. Contrails block not only the incoming solar radiation (reducing the day temperature) but also the outgoing long wave radiation (increasing the night temperature). The absence of the contrails after 9/11 resulted in a 1.1 °C increase in the daily temperature range.

Understanding the net effect of clouds remains an important challenge. Despite the need for a more comprehensive study over time and space, this three-day experiment helped our understanding of the effects of contrails. It also suggested the need to account for this ongoing tug of war between emissions (greenhouse gases) that warm the earth and emissions (smoke and particulate matter) that cool the earth.

The cooling effect caused by aerosols formed the hypothesis of the nuclear winter theory. In 1982, the nuclear winter hypothesis advanced by Carl Sagan, an astrophysicist at Cornell University, caused a debate as intense as the expected firestorms that would start from a nuclear exchange. The theory was simple: fire tornadoes produced during a nuclear war in the northern hemisphere start massive fires.[18] Much of the earth subsequently plunges into darkness as the emissions produced from the nuclear explosions and the fires reduce the solar radiation and initiate global cooling.

The idea of starting forest fires from bombs did not escape the Japanese during the Second World War. Without any long-range aircraft like the B-29, the Japanese relied on an ingenious method to bomb the mainland of the United States. Large paper balloons riding a fast flowing band of air current in the upper troposphere called the jet stream would carry the bombs. The balloons were timed to discharge their bombs over the United States with the expectation that they would ignite numerous forest fires. The balloon bomb attack had the potential to start many firestorms had the Japanese not launched most of the balloons during the winter (November 1944 to April 1945). Of the 9,300 balloon bombs released, an estimated 1,000 landed randomly throughout North America and resulted in two small brush fires.[19]

On August 8, 1974, the New York Times published an article written by Harold Schmeck called "Climate Changes Endanger World's Food Output."[20] Schmeck reported that some scientists were projecting more extreme weather that would threaten the major agricultural areas around the world. Concerns about food shortages in North America focused attention on climate change. Newsweek followed with an alarming story published on April 28, 1975 called "The Cooling World."[21] Writer Peter Gwynne included in his article a graph produced by the National Oceanic and Atmospheric Administration (NOAA) showing the average temperature change since about 1885 to 1970. From 1945 to 1968, NOAA reported a 0.3 °C drop in temperature in the northern hemisphere. "The evidence in support of these predictions has now begun to accumulate so massively that meteorologists are hard pressed to keep up with it," reported Gwynne.[22] Yet, surprisingly, Gwynne included a quote from the 1975 NOAA report that cautions his own claim: "Our knowledge of the mechanisms of climatic change is at least as fragmentary as our data. Not only are the basic scientific questions largely unanswered, but in many cases we do not yet know enough to pose the key questions."[23]

Skeptics argued that the earth was cooling when in fact it was not. A scan of the literature indicates there was no scientific consensus on global cooling. Earlier worries about global cooling and the nuclear winter theory fuelled concerns about smoke. Understanding the likelihood of a nuclear winter event requires an understanding of smoke behaviour. How much smoke is produced? How high is it injected into the atmosphere? Where does it travel and how long does it linger before rain washes it back to earth? Designing experiments to obtain relevant data, validate models, and test the nuclear winter theory is not easy. Prescribed burns and wildfires, therefore, provide opportunities to study what could happen after a nuclear holocaust.

In 1985, Brian Stocks, head of the Canadian Forest Service Fire Research Program at the Great Lakes Forest Research Centre in Sault Ste. Marie, Ontario, invited university and government scientists to observe a prescribed burn near the Town of Chapleau in northern Ontario.[24] Included on the invitation list were officials from the United States Defense Nuclear Agency. The Chapleau prescribed burn was no ordinary controlled fire. The Ontario Ministry of Natural Resources felled and crushed approximately 800 ha of balsam fir trees that had been killed by the spruce budworm, resulting in a large fuel load of dead tree trunks, branches, and needles. The burn objective was to consume this surface fuel layer and prepare the site for the establishment of a new stand of trees.[25] This required a high-intensity fire.

The Chapleau prescribed burn was a successful groundbreaking experiment to observe the behaviour of smoke under real conditions. It became known as a "nuclear winter test fire." Controlled fires provide the opportunity for scientists to quantify the weather conditions and fuel loads before the fire starts. The burn unit can also be fully instrumented to record data such as temperature and rate of spread during the burn. However, few measurements occurred that day. The scientists simply observed a spectacular grandeur of power. Never before had a controlled burn been conducted at this size and intensity for research purposes. The images seared a lasting impression for the visiting scientists.

As interest in the nuclear winter theory waned, smoke chemistry became the next issue to tackle. From 1987 to 1991, the Canadian Forest Service collaborated with the Ontario Ministry of Natural Resources to instrument large prescribed burns in northeast Ontario. This work spawned long-term research linkages with several agencies in the United States, and in particular the National Aeronautics and Space Administration (NASA).

In 2008, Brian Stocks, then retired from the Canadian Forest Service and working as a consultant, assisted university and NASA scientists from the United States to track smoke from fires from June 26 to July 14 in Canada's boreal forest in a large, complex research study called Arctic Research of the Composition of the Troposphere from Aircraft and Satellites (ARCTAS). This $24-million NASA-led international initiative is the largest airborne experiment to study the role of atmospheric pollutants and, in particular, emissions from boreal forest wildfires called pyrogenic black carbon (soot) and their contribution to the rapid warming of the Arctic environment. The observed changes in the northern environment are warnings of what climate change may bring to other regions.

In the summer of 2008, more than 150 scientists and support staff from around the world met at the Canadian Armed Forces Base at Cold Lake, Alberta, and waited excitedly for the boreal forest fires and smoke to arrive.[26] On July 2, 2008, Saskatchewan had 45 reported active fires. A cold front started to move through the province on July 4, creating blowup conditions for many of the northern fires burning out of control in Saskatchewan's fire management observation zone. The chase that began on July 5 was well orchestrated. The National Aeronautical and Space Administration dispatched their DC-8 and P-3B aircraft based at Cold Lake and their B-200 aircraft based at Yellowknife, Northwest Territories to track the fire emissions. Satellite imagery and instruments carried by balloons were released from ground stations. As the NASA B-200 aircraft spiraled downward from the top of one of the smoke plumes, the Cloud-Aerosol Lidar and Infrared Pathfinder Satellite Observation (CALIPSO) satellite tracked the plume from above in synchrony. The researchers also tracked smoke drifting into Canada from fires in California and Siberia. The data collected from the ARCTAS project, including fuels loads and fuel consumption, help to validate computer models that predict emissions and their trajectories. Concerns about changing climate in the Arctic were supported by observations of forest fire smoke frequently flowing to the Arctic.

Although more difficult to document, wildfires hold important clues for scientists who study the impacts of smoke injected into the stratosphere. Craig Chandler, retired director of fire research for the United States Forest Service, completed a special study on the nuclear winter theory for the United States Department of Defense. In May 1984, he contacted Peter Murphy at the University of Alberta to obtain information about the 1950 Chinchaga Firestorm.[27] He was also interested in the July and August 1982 wildfires in the Yukon that burned approximately two million hectares. Using satellite imagery, NASA scientists tracked the smoke from these fires to England. In Chandler's letter to Murphy, he noted, "The idea of a 50 °C drop in temperature probably doesn't phase you hardy Canadians, but it seems to have scared [the] hell out of the sun worshippers in the lower 48."[28]

The sun worshippers were not just military officials and scholars trying to prove or disprove the nuclear winter theory. With the help of a well-funded marketing machine, the frigid horror of a nuclear winter scenario became branded in the public's psyche. A special issue of *Ambio*, published in 1982 by the Royal Swedish Academy of Sciences, was particularly influential in creating the image of an apocalyptic nuclear winter. The issue assessed the possible human and ecological consequences of a nuclear war.[29] One of the articles in this issue, titled "The Atmosphere after a Nuclear War: Twilight at Noon," planted the seed for the nuclear winter theory.[30] Authors Paul Crutzen and John Birks suggested that vast areas of forests would burn, producing a thick smoke layer that would drastically reduce the amount of solar radiation reaching the earth.

The special issue of *Ambio* was edited by Jeannie Peterson. Her husband, Russell Peterson, who was president of the Audubon Society, an American non-profit environmental organization dedicated to ecosystem conservation, approached the Rockefeller Family Fund and the Henry P. Kendall Foundation to increase their awareness of Crutzen and Birks's study. Funding soon flowed from these two organizations to support a quickly assembled team of over 100 scientists. A scientific advisory board led by Carl Sagan managed the project.[31]

Sagan argued for more critical thinking. The strategy of nuclear deterrence requires a country to maintain a sufficient quantity of nuclear warheads to annihilate another country if they choose to attack first. Sagan suggested reducing the nuclear stockpiles to a level that would not trigger a nuclear winter. Sagan's colleagues intensely criticized this policy recommendation because the science was not credible. For this reason, Sagan's nomination in 1992 to the National Academy of Sciences was rejected. Science is tentative; good science stays, and bad science is purged. There is consensus in the science community that devastating climatic changes can be triggered by a nuclear war, but the notion of a global deep freeze is no longer a hot topic.

When temperature inversions suppress vertical mixing of the air, as was the case during Black Sunday, trapped aerosols, such as smoke, can reside for extended periods of time and travel great distances. The smoke from the Chinchaga Firestorm, and other fires in the area, remained airborne for over seven days.

Aerosols and Human Health

Aerosols impairing air quality and human health are called pollutants. They include: ground level ozone (O_3), particulate matter (PM), sulphur dioxide (SO_2), carbon monoxide (CO), methane (CH_4), nitrogen oxides (NO_x), volatile organic compounds (VOCs), polycyclic aromatic hydrocarbons (PAHs), hydrogen sulphide (H_2S), sulphates, nitrates, and trace minerals. Exposure to pollutants damages the respiratory system, which can then trigger a cascade of secondary injuries to the cardiovascular and reproductive systems.[32] Any increase in the amount of particulate matter will increase the health risk to the public to some extent.[33] Young and old people and anyone with respiratory and cardiovascular problems are the most susceptible to even small increases in the concentration of fine particulate matter.

Carbon dioxide and water vapour account for 90% of the emissions from wildfires.[34] Many of the PAHs in smoke are carcinogenic. Benzene, in particular, is very carcinogenic. Although smoke is a toxic cocktail of thousands of compounds, the greatest health hazard of smoke is the fine particulate matter.[35]

In 1905, the term *smog* was coined to describe the mixture of smoke and dense fog that occasionally lingered over London and other cities in Great Britain.[36] Today, *smog* refers to a mixture of ozone and aerosols. Ozone is an odorless, colourless gas created when hydrocarbons, nitrogen oxides, and water vapour react chemically in the presence of sunlight.[37] This is why smog is more prevalent during summer. Smog can form as a thick stagnant blanket over communities. It can affect the health of children, elderly people, and anyone with cardiorespiratory diseases such as congestive heart failure, angina, emphysema, bronchitis, and asthma. Smog affects the health of one out of three people in the United States.[38] The health and social costs associated with smog are staggering. The Ontario Medical Association estimated that smog costs over one billion dollars a year in Ontario alone due to increased hospital admissions and absenteeism from work.[39] The number of residents admitted to hospitals in Ontario with respiratory problems is forecasted to increase to 24,000 by 2026.[40] An estimated 10,000 Ontario residents will die prematurely because of smog events.[41] Most of these people will already be suffering from cardiorespiratory diseases. Smog is the grim reaper that cuts their lives short.

In 2008, the Canadian Medical Association published an alarming report titled *No Breathing Room—National Illness Costs of Air Pollution*.[42] They estimated 21,000 Canadians died in 2008 because the air they breathed contained pollutants. Over 90% of these deaths occurred because of chronic and long-term exposure to air pollution. By 2031, the number of deaths will increase to a startling 710,000. An additional 90,000 people will die prematurely from the acute exposure to air pollution. People with cardiovascular disease will account for 42% of the acute premature deaths. Eighty percent of these deaths will strike those over 65 years of age.[43]

When pollutant concentrations reach critical levels, smog alerts and health advisories are usually given. The United States Environmental Protection Agency developed an Air Quality Index (AQI) to help people protect themselves when the air quality deteriorates.[44] An AQI of 201 to

300 represents very unhealthy air quality conditions. Under these conditions, individuals most at risk to exposure are advised to avoid outdoor physical activity. All other people are encouraged to avoid extended exposure and intense activities outdoors. Canada also has an index to report the relative level of pollutants in the air. This scale, called the Air Quality Heath Index ranges from 1 (very low) to 10+ (very high).

Before the official start of summer, the Ontario Ministry of the Environment issued 12 smog alerts for the city of Toronto in June 2005.[45] Record-setting temperatures and a lack of rain provided ideal conditions for smog to persist from June 2 to 14. Most of the small particulates in the smog originated from factories, automobiles, and coal-powered generating stations in the eastern United States. Ontarians also have a dependency on vehicles and power-generating stations, but the last coal power plant, located in Thunder Bay, shut down in 2014. Ontario is now coal free. Nevertheless, given that bad air moves freely across the border, we all should strive to minimize the impact from our activities, which includes reassessing what vehicles we drive and how often we drive them.

Governments and the public around the world became more aware of the lethal effects of smog after an extreme air pollution event in 1952 in London, England. On December 12, a killer smog smothered the city for four days.[46] Smoke from millions of residential and industrial coal fires mixed with natural fog to create a black, poisonous cloud that accounted for 12,000 excess deaths from December 1952 to February 1953. With visibility reduced to one foot, and roads cluttered with abandoned vehicles, Londoners with smog-filled lungs struggled blindly to find their way to a city hospital. Nurses at the Royal London Hospital reported many patients with blackened under garments because the smog penetrated their clothes.

The Great Smog of London in 1952 triggered an ongoing battle for clean air. After the smog event, legislation was quickly passed to phase out the use of residential coal fires. The City of London now has a network of 119 air quality monitoring stations.[47] Air quality standards are now the norm in many countries.

The widespread power outage over eastern North America on August 14, 2003 drastically reduced the emission of pollutants, temporarily resulting in bluer, clearer, and cleaner skies. The response was unexpected. Aerosol levels dropped from one-third to one-quarter of normal levels, and ozone concentrations were reduced by one-half.[48]

In Canada, forest fires account for approximately one-third of particulate matter emissions.[49] Forest fires can generate high concentrations of pollutants that can impair air quality for considerable distances downwind. Abrupt changes in air quality often occur with little warning. Smoke from forest fires is a silent killer. Each year, smoke events around the world cause the premature death of people suffering from cardiorespiratory diseases. When smoke enters communities, hospital admissions and premature deaths increase proportionally with the duration and concentration of the smoke. And increasing evidence suggests that the related health costs are significant. The Canadian Medical Association projects the economic cost of air pollution to increase from approximately $10 billion in 2008 to over $300 billion by 2031.[50] Worldwide, smoke from landscape fires contributed to an average of 339,000 deaths each year from 1997 to 2006.[51]

Based on simulation modelling, graduate student Robyn Rittmaster, and her supervisor Vic Adamowicz at the University of Alberta, modelled the economic impact of the health effects from forest fires using the 2001 Chisholm Fire in Alberta as a case study.[52] The estimated total health costs associated with the Chisholm smoke event on May 24, 2001 were approximately four million dollars. In the capital city of Edmonton, located 160 km southeast from the fire, the one-day smoke event increased the average PM2.5 level to 55 µg/m³. At this level, individuals sensitive to air pollution may experience health effects. The hourly PM2.5 level reached 261 µg/m³. This level is considered very unhealthy for the general population. According to the Canada-Wide Standards for air quality, the PM2.5 levels averaged over a 24-hour period should not exceed 30 µg/m³. Even in the city of Red Deer, located 345 km south of the fire, the air quality exceeded the Canada-Wide Standard for levels of PM2.5.

The health cost estimated by the University of Alberta researchers for the Chisholm Fire suggest the air quality impacts were significant when compared to the other impacts reported in the 2001 Chisholm Fire Review Committee Final Report.[53] The researchers used a prediction model developed collaboratively by Environment Canada and Health Canada called the Air Quality Valuation Model (AQVM). Because epidemiology studies are time-consuming and expensive, the AVQM provides an alternate approach to predict health outcomes and economic valuation estimates using atmospheric concentrations of various pollutants, including PM2.5. The concentration-response function, an input in the model, is based on an underlying database of impacts derived from pollution exposure studies.

There are limitations to using the AQVM. Increased premature mortality accounted for approximately 95% of the health costs of the Chisholm Fire. However, there is considerable uncertainty regarding the exposure response relationships for mortality. As well, estimating the monetary value of an individual who dies from a smoke event is difficult. The value of a statistical life, referred to as VSL, is not an estimate of the intrinsic value of a life. It is an estimate based on a willingness of a group to pay for a reduction in the risk of mortality. Not surprisingly, VSL values vary considerably.

In 1999, a research team sponsored by the United Nations Environment Programme (UNEP) studied a three-kilometre thick toxic cocktail of airborne particulates and chemicals that blanketed south Asia in 1998, which was the result of biomass burning, including forest fires and agricultural and subsistence burning.[54] Known as the "brown cloud," this pollution contributed to the premature deaths of 500,000 people in India alone. The UNEP study of the impact of the brown cloud surprised many of the over 200 scientists who were participating in the study. Of particular interest was the unexpected distance the pollutants travelled. Blocks of pollution such as the brown cloud can travel half way around the earth in one week. Local pollution problems quickly become regional and global pollution problems.

In August 2005, a brown cloud again shrouded parts of Southeast Asia because of drifting smoke from forest fires in Indonesia.[55] Malaysia subsequently declared a state of emergency in several areas outside Kuala Lumpur. The World Health Organization estimates that air pollution causes more than two million deaths annually.[56] As well as the premature deaths from air pollution, developing countries in Southeast Asia suffer from diseases sensitive to a warming climate. Increasing temperatures favour mosquito-borne illnesses such as Dengue fever and malaria.

Events like the brown cloud reduce the amount of solar energy reaching the surface of the earth by up to 15%.[57] This reduces the amount of evaporation from the oceans, which in turn influences the amount of precipitation.

Since the smoke from the 1950 Chinchaga Firestorm in western Canada remained at high elevations, the air quality at ground level remained unchanged. However, forest fire smoke entering United States from Canada does not always travel and stay at high levels. On July 2, 2002, a severe thunderstorm left a trail of 45 new lightning fire starts in northern Quebec. Several of these fires southeast of James Bay quickly raged out of control, forcing the evacuation of approximately 630 residents in the Cree communities of Nemaska and Chisasibi.[58]

In Quebec, fires occurring above the 52nd parallel of latitude are often left to burn. This zone is called the modified suppression zone. Embedded within this zone are several community protection zones. All fires south of the 52nd parallel are aggressively suppressed. This area is called the full suppression zone.

From July 6 to 8, 2002, the smoke from the fires in the modified suppression zone travelled south into the eastern United States as far south as Virginia. Stan Coloff, fire coordinator with the United States Geological Survey based in Washington, DC, was sailing on July 7 when the smoke arrived. He recalled looking north, watching what appeared to be an approaching storm. However, the weather reports he checked prior to setting sail indicated it was supposed to be a clear, sunny day. How could the forecast miss this large storm?[59]

The blanket of smoke covered parts of New York, New Hampshire, Vermont, and Massachusetts. In Pennsylvania, the Department of Environmental Protection issued a health advisory for 20 counties; and in New York, the Health Department issued a state-wide health advisory.[60] This smoke event struck when the most damage could be done—a warm summer weekend busy with outdoor enthusiasts, both young and old. On July 9, the winds changed direction and carried the smoke eastward toward the Atlantic provinces.

Health advisories were also issued in Ontario and Quebec. The measured concentration of smoke particulates in Montreal was three times the limit of acceptability.[61] In Ontario, the highest concentration of particulate matter (24-hour average) occurred in Peterborough (71.8 μg/m³) and Ottawa (70.4 μg/m³) on July 6.[62] The 2002 Quebec fires were responsible for 84% of the total provincial PM2.5 emissions for that year.[63]

Three monitoring sites in Baltimore, Maryland, measured maximum particulate matter concentrations of 199, 645, and 590 μg/m³. The estimated 24-hour average of 86 μg/m³ on July 7 exceeded the national standard in the United States (65 μg/m³).[64] Even though the smoke travelled a distance of approximately 1,400 km, the concentration remained high enough for health advisories to be issued. A particulate concentration level of 645 μg/m³ is considered hazardous.

On May 25, 2010, lightning struck again. Eight of the 54 active fires north of Trois-Rivières, Quebec, burned out of control on May 30, 2010. Smoke from these fires travelled southeast impacting the air quality at Quebec City and Ottawa, and the cities of Portland (Maine), Manchester (New Hampshire), and Boston (Massachusetts).[65] Despite the smoke event, many residents in northeast United States decided to stay outdoors to celebrate their Memorial Day holiday on May 31.

Staying indoors is not necessarily the best strategy when a health advisory is issued. According to researchers at the University of British Columbia, up to 70% of the smoke will enter homes if they are not properly sealed.[66] They recommend the use of HEPA air filters to remove the fine particulates and the establishment of clean air shelters.

Smoke events from forest fires occur every year around the world. In 1997, fires in Indonesia burned deep in drained peatlands and released a staggering surge of carbon into the atmosphere. The forest and peat fires in Indonesia alone unlocked an estimated 0.8 to 2.6 Gt of carbon or the equivalent of 40% of the annual global carbon emissions from fossil fuels.[67] On July 31, 2002, smoke from over 100 forest fires burning out of control shrouded the city of Moscow.[68] These fires, too, burned stubbornly in drained peatlands. Landmarks, such as the Christ the Saviour Cathedral, became eerie shadows until rain arrived on the weekend. The next year, wildfires in Siberia consumed 22 million hectares. These fires released an estimated 2,250 million tons of carbon into the atmosphere, primarily in the form of carbon dioxide and carbon monoxide.[69] Since carbon dioxide is a major contributor to global warming, wildfire emissions cannot be ignored. Merritt Turetsky at the University of Guelph is concerned that the world's peatlands may switch from carbon sink to carbon source if their water tables become lowered due to land-use practices (drainage) and climate change. Her research suggests these altered peatland ecosystems may release a ninefold increase in carbon if burned.[70]

In 2004, smoke from large fires burning from mid-June to mid-July in Alaska and Yukon travelled south into United States as far as Florida.[71] Satellites tracked the smoke plumes from these fires to help scientists better understand and model how smoke aerosols are transported around the globe. Scientists at the Jet Propulsion Laboratory in Pasadena, California, used images taken from the Multi-angle Imaging SpectroRadiometer (MISR) on NASA's Terra satellite to differentiate smoke from clouds and to estimate the height and depth of smoke plumes.[72] The 2004 forest fires in Yukon accounted for 10% of the emissions in Canada for that year.[73]

In late September 2006, around 300 fires arrived and splashed an already autumn red forested landscape in northwest Ontario with more red.[74] Community protection became the top priority. With many of the seasonal firefighters already back in school and unseasonably dry conditions, the fires were allowed to burn freely. Thick acrid smoke from these fires travelled over eastern Canada.

Where the smoke from the 1950 Chinchaga Firestorm travelled did not concern Forest Ranger Frank LaFoy. The relationship between smoke exposure and health was not well understood in 1950. LaFoy, too, focused on the protection of life and property. Fire, not smoke, threatened his community.

Smoke filled the skies from June to September. Dominion land surveyor R.W. Thistlethwaite and his survey crew coped with severe smoke conditions while travelling and surveying the British Columbia–Alberta boundary in 1950. The prolonged exposure to the smoke during the entire fire season would have impacted the health of the Keg River area residents to some extent. Some individuals may have only experienced eye irritation and laboured breathing. Others may have had more severe respiratory and cardiovascular symptoms, but Dr. Mary Jackson did not report any deaths as a result of the reduced air quality.

Emissions from wildfires are estimated by multiplying the area burned (ha) by the amount of fuel consumed (kg/ha) during both the flaming and smoldering phases of combustion. The Canadian Fire Behaviour Prediction (FBP) System outputs estimates of fuel consumption (kg/m² where 1 ha = 10,000 m²).[75] Obtaining accurate estimates of total fuel consumption is a challenge because the fire environment changes over time and space. The variability in fuel characteristics, topography, and weather creates variability in fire behaviour and effects across the landscape.

Once an estimate is made of the total amount of fuel consumed by the fire, the amount of a particular emission released can be estimated by using a multiplication emission factor in grams per kilogram of fuel consumed. For example, the emission factor for PM2.5 is 14.0 +/- 9.1 (grams of particulate matter per kg of dry matter consumed).[76]

In the late 1990s, the Canadian Forest Service began development of a new boreal fire effects model called BORFIRE (the next generation model was renamed CANFIRE). This model uses a new forest floor fuel consumption algorithm and pre-fire fuel load data from a National Carbon Budget Model.[77] CANFIRE produces higher fuel consumption and emission rates than the FBP System, which generally underestimates fuel consumption.

If Frank LaFoy worked today as a forest ranger, his forest ranger tool-box would include a smoke forecasting system and models to determine the trajectory of the smoke plume and concentration of particulates. These tools allow the public to be warned in advance of a smoke event and the associated health risks. Fire managers use these same tools to mitigate the impact of the smoke on the public from prescribed burns.

LaFoy also did not have the benefit of using satellite imagery. On October 4, 1957, the Soviet Union successfully launched the first satellite into space. Thousands of satellites loaded with specialized sensors and instruments now orbit the earth, collecting information we rely upon daily. Some of these satellites collect specific information about how much smoke there is and who is making it. The Moderate Resolution Imaging Spectroradiometer and Multi-angle Imaging Spectroradiometer onboard the Terra satellite measure Aerosol Optical Depth (AOD). The CALIPSO and a network of ground-based sun photometers called

AERONET (Aerosol Robotic Network) are used to establish a relationship between AOD and PM2.5 concentrations.

Monitoring who owns and releases carbon is increasingly becoming important. In Canada, the amount of carbon emitted as carbon dioxide, carbon monoxide, and methane by fires during extreme years can approach levels of industrial carbon emissions.[78] Actions in one country can affect other countries. If fires in the boreal forest release more carbon than they capture, fire managers become carbon players on the global stage.

SIX

THE BIG WIND

The Chinchaga River Fire was one of the large fires that burned during the 1950 Chinchaga Firestorm in western Canada. Although it burned from June 1 to September 30, 81% of the modelled total linear fire spread occurred during 15 days (12.3%) (see Table 6.1).[1] Forest fires in the boreal forest burned historically through the entire fire season. In the absence of fire suppression, these fires spread slowly during long periods of time and rapidly during short periods when wind events occurred. This pattern continued until the arrival of a fire-ending rain or snow event. Some fires overwintered in a smoldering stage and awoke again in the spring.

The fire growth simulation modelling of the Chinchaga River Fire identified five short periods when major fire activity occurred (Figure 6.1,

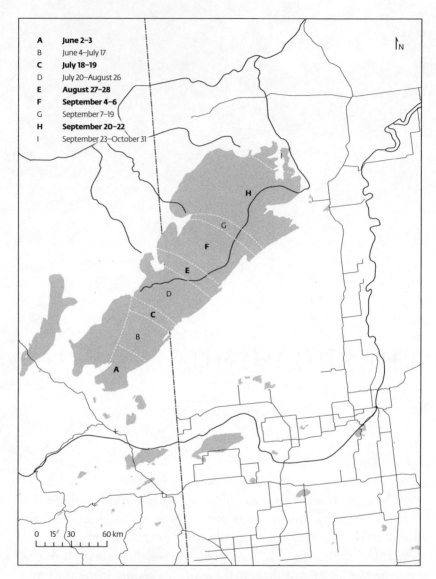

A	**June 2–3**
B	June 4–July 17
C	**July 18–19**
D	July 20–August 26
E	**August 27–28**
F	**September 4–6**
G	September 7–19
H	**September 20–22**
I	September 23–October 31

0   15   30   60 km

FIGURE 6.1   Fire growth simulation of the 1950 Chinchaga River Fire. Source: Murphy and Tymstra, "The 1950 Chinchaga River Fire." Bold dates correspond to the five short periods when major fire activity occurred.

TABLE 6.1   Weather, fire weather indices, and modelled linear spread distances for the 1950 Chinchaga River Fire. Data in this table are based on weather observations taken at 1230h at the Fort St. John Airport weather station in British Columbia. Grey sections correspond to the five short periods when major fire activity occurred.

Period	Maximum Temperature, °C	Minimum RH, %	Maximum 10 m Wind Speed, km/h	Initial Spread Index (ISI)	Buildup Index (BUI)	Period Spread, km	Cumulative Spread, km
June 2	22.0	21	48	27	47	21	21
June 3	21.0	44	42	30	51	13	34
June 4–9	–	–	–	–	–	1	35
June 10	26.8	27	35	15	73	8	43
June 11–July 17	–	–	–	–	–	20	63
July 18	25.0	40	45	29	30	7	70
July 19	18.9	47	40	20	33	4	74
July 20–29	–	–	–	–	–	6	80
July 30	19.1	37	48	15	51	9	89
July 31–Aug. 26	–	–	–	–	–	19	108
Aug. 27	19.4	37	39	17	64	6	114
Aug. 28	17.2	53	45	10	66	5	119
Aug. 29	16.8	47	37	14	68	3	122
Aug. 30–Sept. 3	–	–	–	–	–	1	123
Sept. 4	18.7	37	58	21	71	18	141
Sept. 5	19.3	40	52	18	74	13	154
Sept. 6	15.6	43	48	15	76	5	159
Sept. 7–19	–	–	–	–	–	7	166
Sept. 20	22.1	47	64	44	104	35	201
Sept. 21	24.1	45	29	11	107	2	203
Sept. 22	26.1	26	55	15	111	26	229
Sept. 23–Oct. 31	–	–	–	–	–	16	245

Table 6.1).[2] The greatest fire spread occurred from September 20 to 24. On one day alone, September 20, the fire raced an estimated 35 km towards the village of Keg River.[3] Pilot Johnny Bourassa flew over the Chinchaga River Fire on November 11 and reported seeing little fire activity except for the still smoldering muskeg areas.[4]

Peter Murphy and I obtained weather observations from the Atmospheric Environment Service for 1949 and 1950, for Fort Nelson,

TABLE 6.2 Summary of the observed surface weather for six meteorological stations, September 20–24, 1950. (RH = relative humidity; WS = wind speed; WD = wind direction)

Weather Station	Sept. 20				Sept. 21			
	Temp °C Max/Obs	RH %	WS km/h	WD	Temp °C Max/Obs	RH %	WS km/h	WD
Fort St. John Airport Hourly (1230h)	22.1 / 21.0	47	52	SW	23.9 / 22.0	52	24	SW
Fairview Hourly (1230h)	26.1 / 23.0	40	16	SW	25.0 / 21.5	48	6	SW
Fort Nelson Airport Hourly (1230h)	25.5 / 14.5	95	8	WSW	21.6 / 16.5	52	6	NE
Beatton River Airport Synoptic (1730h)	23.7 / 21.6	32	39	WSW	24.5 / 23.3	39	18	SW
Keg River Synoptic (1730h)	25.5 / 23.3	59	32	SW	23.3 / 22.2	62	10	E
Peace River Airport Hourly (° 1815h; † 1915h)	27.2 / 20.0°	46	16	S	25.5 / 25.0†	–	13	SW

Fort St. John, and Beatton River in British Columbia, and Fairview, Peace River, and Keg River in Alberta (Table 6.2). This historical data were used to calculate daily and hourly outputs of the Fire Weather Index (FWI) Subsystem of the Canadian Forest Fire Danger Rating System.[5]

The FWI Subsystem of the Canadian Forest Fire Danger Rating System consists of three fuel moisture codes—fine fuel moisture code (FFMC), duff moisture code (DMC), and drought code (DC)—and three fire behaviour indices—buildup index (BUI), initial spread index (ISI), and fire weather index (FWI).[6] The FWI Subsystem requires consecutive daily weather observations taken at noon (1200h LST or 1300h DST) to calculate the codes and indices. FFMC is a relative measure of the moisture content of the surface fine fuels. DMC is a relative measure of the moisture content of the duff layer (seven centimetres in depth), and DC is a relative measure of the moisture content of large diameter surface fuels and the deep duff layer.

The 1950 starting DC was adjusted using the 1949 season-ending DC value and the 1949–50 overwinter precipitation (total water equivalent). DMC and DC are calculated as daily codes, whereas, FFMC can be calculated as

TABLE 6.2 *(continued)*

	Sept. 22				Sept. 23				Sept. 24		
Temp °C Max/Obs	RH %	WS km/h	WD	Temp °C Max/Obs	RH %	WS km/h	WD	Temp °C Max/Obs	RH %	WS km/h	WD
26.1 / 25.0	35	23	S	19.5 / 18.5	55	36	SW	17.8 / 14.5	49	24	S
28.3 / 22.0	48	5	SE	24.4 / 17.0	53	5	NW	15.5 / 11.0	89	8	S
29.4 / 24.0	45	10	S	25.0 / 14.0	62	6	NE	20.0 / 18.0	32	19	SW
23.7 / 22.0	23	35	SW	19.1 / 18.8	37	32	SSW	17.2 / 15.6	37	29	SW
30.5 / 25.5	47	24	SW	25.5 / 23.3	59	21	SW	23.3 / 18.8	49	26	SW
27.2 / 22.7†	49	6	S	– / 17.7°	–	10	S	22.2 / –	–	–	–

a daily or hourly code. Likewise, with the fire behaviour indices, the BUI is calculated daily, whereas ISI and FWI can be calculated daily or hourly.

The calculation of the 1949 FWI codes and indices began on the third day after the mean date of last snow cover. These dates are April 29 for Fort Nelson, April 25 for Fairview, and April 24 for Fort St. John. The calculation of the 1950 FWI codes and indices for Fort Nelson and Fairview also began on the third day after the mean date of last snow cover. These dates are April 29 for Fort Nelson and April 25 for Fairview. The 1950 FWI calculations for Fort St. John began on April 29, the third day after the observed (i.e., actual) date of snow-free cover. Standard starting values used for the 1949 FWI calculations for Fort St. John are FFMC = 85, DMC = 6, and DC = 15. The starting values for the 1950 FWI calculations for Fort St. John are FFMC = 85, DMC = 25, and adjusted DC = 261.[7] The starting fuel moisture codes give an indication of the hazard conditions at the start of the fire season. The drought code was adjusted based on the ending DC value for the previous season and the total amount of overwinter precipitation. A DC value of 261 suggests drought conditions occurred in the spring. In Alberta, a DC value of 261 is considered very high.

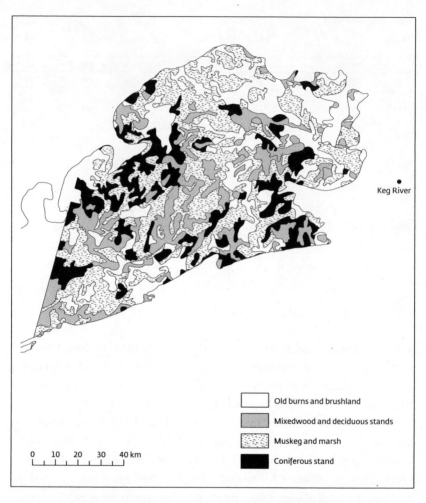

Old burns and brushland

Mixedwood and deciduous stands

Muskeg and marsh

Coniferous stand

0 10 20 30 40 km

Keg River

FIGURE 6.2   Shown are pre-burn vegetation classes for the area covered by the Chinchaga River Fire in Alberta. Data for BC are not available. 1957 Alberta Forest Classification Map.

The Fort St. John weather was used in the simulation shown in Figure 6.1. The linear fire spread distance was calculated using the Fire Behaviour Prediction Subsystem of the Canadian Forest Fire Danger Rating System and applying the rate of spread equation for Fuel Type C-3 (mature jack or lodgepole pine).[8] The pre-burn forest cover consisted primarily of coniferous, mixedwood, and deciduous stands up to 18.3 m (60 feet) in height (Figure 6.2). Daily spread distances using daily

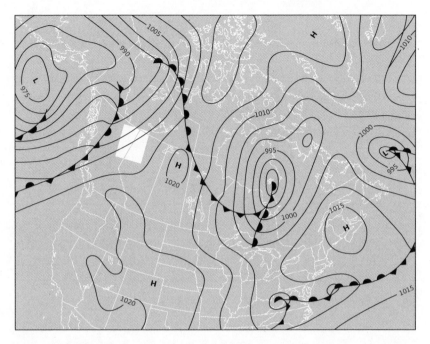

FIGURE 6.3   Synoptic weather map for the northern hemisphere, September 22, 1950, 0530h. The fire area is located at the western edge of a high pressure system, and at the eastern edge of a low pressure system.

weather input were calculated for June 1 to September 30, 1950. The rates of spread for eight short periods totalling 15 days were recalculated using hourly weather data.

During the months of August and September, the fuel moisture codes climbed quickly because of the presence of a blocking high (strong, nearly stationary region of high atmospheric pressure at the surface and aloft). On September 20, the DC reached 549, which indicated extreme drought conditions.[9] The buildup and breakdown of blocking highs and the passage of cold fronts cause many large fire spread events in the boreal forest. A blocking high from September 5 to 23 resulted in below average precipitation and unusually high temperatures (Figure 6.3).[10] On September 22, Keg River reported a temperature of 30.5 °C.[11] Below average precipitation occurred not only in September but also in the preceding month. Keg River received below average precipitation during

TABLE 6.3 Monthly 1950 precipitation values (mm) for weather stations in Fort St. John and Fort Nelson, BC, and Fairview and Keg River, AB. Normal precipitation values (1941–1970) shown in grey.

Month	Fort St. John		Fort Nelson		Fairview		Keg River	
January	9.9	33.3	10.4	26.4	14.2	27.7	10.9	21.8
February	38.1	28.0	34.8	24.4	55.9	29.2	36.8	21.6
March	7.1	27.2	13.2	24.9	7.6	19.1	2.5	20.3
April	30.9	23.4	9.9	21.6	26.4	19.1	27.9	19.8
May	42.9	33.0	10.9	37.6	38.6	30.0	66.5	35.6
June	57.6	61.7	24.1	64.3	52.5	64.8	20.3	53.3
July	57.6	62.7	41.9	74.7	74.4	52.8	13.0	59.4
August	29.2	53.6	34.5	55.4	49.2	51.3	41.1	51.5
September	9.9	32.5	7.1	38.1	24.1	38.1	47.2	44.7
October	16.0	29.7	30.7	25.6	22.4	24.9	15.7	21.3
November	55.8	32.0	61.4	26.7	70.6	33.8	48.8	27.4
December	19.8	32.8	36.3	26.0	21.6	31.5	27.9	26.2

the months of June, July, and August.[12] Only 11.4 mm (0.45 inches) of precipitation occurred from August 16 to September 24.[13] Fort St. John received 54% of the normal precipitation (1941–1970) during August and only 30% during September based on the monthly averages for the period 1941 to 1970.[14] (See Table 6.3.)

Each day, as forest fuels dry under the influence of a blocking high, progressively larger diameter fuels and deeper organic layers become available for consumption by a wildfire. The fire hazard thus climbs. The air mass associated with a blocking high (also referred to as a ridge) is relatively stable, resulting in no or little lightning. The surface winds are also light. Even with active fires, or new human-caused fire starts, fire management agencies usually encounter few problems because the light winds result in low fire rates of spread. Fire managers know the blocking high will eventually collapse, and the winds will return. Frank LaFoy reported strong southwest winds at Keg River on September 20.[15] These winds increased in velocity over the next two days. When Forest Ranger Mitchell completed the final fire report for the Whisp Fire in British Columbia, he also noted strong winds on September 20: "Fire burned over most of the slash and did not run out in open country until after the unusually long dry spell in September when the gale [force] winds on

Sept. 20 fanned it up and it raced 30 miles across country."[16] In its path was Dan Hack's homestead. He lost his farm buildings and three horses. Upset with the Forest Service for failing to fight the fire, Hack met with the assistant district forester to seek reimbursement. Since Hack's farm was outside the firefighting zone, the district forester advised him nothing could be done.[17]

British Columbia Forest Ranger Mitchell and Alberta Forest Ranger Sears made important independent observations about the location of the fire. Sears estimated the fire was still in British Columbia when he took a compass bearing (estimated to have occurred on September 7). In Mitchell's fire report, Dan Hack's homestead reportedly burned on September 20. This suggests the majority of the total spread distance of the Chinchaga River Fire occurred during September, and in particular, during the September 20 to 22 period. Fire growth simulation modelling of the Chinchaga River Fire corroborates Mitchell's and Sears's observations. From June 2 to September 19, the Chinchaga River Fire spread an estimated 166 km. This distance, which represents 68% of the total spread distance, positions the fire near the British Columbia–Alberta border. The fire then raced 79 km during the three-day period from September 20 to 22.[18]

According to Frank LaFoy, the "big wind" hit Keg River on September 22.[19] That day, wind gusts reached 80 km/h and the relative humidity dropped to 26% at 1430h at Fort St. John, British Columbia.[20] Strong winds developed because a large upper low pressure system moving eastward rammed into the west edge of the blocking high (Figure 6.3). Winds from both systems converged over the fire area.

The atmosphere is in constant motion, both horizontally and vertically, in an effort to reach an equilibrium state. For three weeks, the blocking high deflected disturbances that were trying to push it eastward. When the fires made their major runs during the September 20 to 22 period, a conditionally unstable atmosphere occurred over western Canada. On September 24, the ridge of high pressure broke down (Figure 6.4).[21]

In the northern hemisphere, winds circulate clockwise and outward around the centre of a high pressure system (anticyclone) and counter-clockwise and inward around the centre of a low pressure

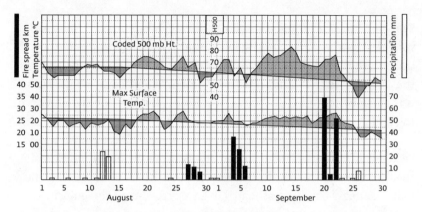

FIGURE 6.4   Chart of 500 mb height/maximum temperature anomalies with daily precipitation amounts at Fort St. John, BC, June 1–September 30, 1950. The projected spread distances are also shown. Compiled by B. Janz, and included in Murphy and Tymstra, "The 1950 Chinchaga River Fire."

system (cyclone). The convergence of both systems over the Chinchaga River Valley resulted in high pressure gradients and, hence, very strong winds. The stronger the gradient or difference in pressure, the stronger is the wind speed. The topography also helped to funnel the winds.

The combined effect of the strong winds and the topography is evident in the shape of the fire perimeter on its east flank. A sharp line following the top of the Clear Hills is visible on the Earth Resources Technology Satellite (ERTS-1) imagery. The location of the south and east perimeter, as mapped by Ron Hammerstedt from the Alberta Forest Service, was confirmed by B.H. Johnson who transferred from Peace River to Keg River as the district forest ranger in 1955. He knew the boundaries of the Chinchaga River Fire well, and, in 1983, he drew, from memory, a rough boundary of the fire on a map for Peter Murphy.[22] The south and east boundaries matched the final perimeter as drawn by Hammerstedt.

The breakdown of blocking highs is common in the boreal forest.[23] A blocking high establishes itself, fuels dry, and then the ridge of high pressure breaks down and is replaced by a low pressure system. It is the passage of a cold front or the leading edge of a relatively colder air mass that causes havoc for fire managers. Strong winds and lightning occur ahead of and behind the front. If rain accompanies the frontal passage, new fires started by lightning will likely be extinguished. Without rain, new fires arrive. This is called a dry cold front. Cold fronts bring strong winds and also a sudden wind shift, which usually causes more problems than the wind itself. Winds typically change from southwest in front of the front, to west, and then to northwest, behind the front. You can usually tell when a cold front arrives. The temperature drops and you experience a blast of erratic and gusty wind. You can often see the dark leading edge of the air mass as it approaches.

Dominion land surveyor R.W. Thistlethwaite provided detailed accounts of the dry conditions in his survey *Monument Book* and his *Report and Field Journal*.[24] On May 16, 1950, he departed Fort St. John with a crew of 18 men, 25 packhorses, and 5 saddle horses to continue the Alberta–British Columbia Boundary Survey. Trucks hauled the survey party, food, and equipment north for 51 km until the dirt road ended and the Donis Trail began. The survey party followed this trail north to a string of three small lakes draining into the Chinchaga River. The trail headed in a northeast direction towards the Chinchaga River just west of the boundary. After crossing the Chinchaga River, Thistlethwaite continued to follow the trail east and then southeast, until he reached the boundary at Monument 97-5 (north end of Township 97) on June 9 (Figure 6.5). This survey monument was built in 1923. The survey monuments were made by cementing pipe posts with brass caps into the ground or holes chiseled into rock. Each survey monument marked an important survey point on the surface of the earth.

Thistlethwaite's crew encountered smoke haze during their 24-day journey to reach the boundary. This smoke probably came from the Chinchaga Firestorm (Fire 19) burning in British Columbia. From June 10 to 13, Thistlethwaite trained his survey crew by retracing the boundary from Monuments 97-3 to 97-5. Along the way, he met a seismic crew working nearby for Imperial Oil Ltd. The location and timing of Thistlethwaite's encounter with the seismic crew supports LaFoy's claim that smudges left by a seismic crew were additional causal agents.

For the next six days, Thistlethwaite and his crew surveyed and cut a new boundary line from Township 97 to Township 108. Thistlethwaite realized they were fortunate that the summer was extremely dry. Traversing the muskeg areas by horse and foot would not be possible during a wet season.

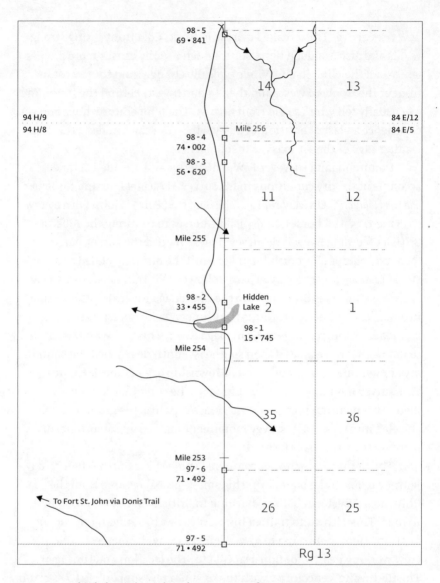

FIGURE 6.5   Thistlethwaite's 1950 boundary survey map from survey Monument 97-5 to 98-5. The trail taken by the survey crew runs north, curving west around Hidden Lake. Source: Thistlethwaite, "Alberta–British Columbia Boundary," in *Monument Book*.

The survey crew crossed two major rivers—Fontas and the Hay. Thistlethwaite described the flow of water in the Fontas River on August 1 as "practically negligible."[25] The appearance of the banks suggested normal water levels of 1.2 m to 1.5 m (4 feet or 5 feet) deep. The Fontas River Valley consisted primarily of spruce with some stands up to 61 cm (24 inches) in diameter.

At its southern boundary crossing, Thistlethwaite estimated the Hay River Valley to be 800 m (2,640 feet) wide and 33.5 m (110 feet) deep—much larger than the Fontas River Valley. When the survey crew crossed the Hay River on September 19, the flow of water was described as "inconsequential" and consisting of "trickles running from pool to pool through the gravel and boulders of the stream bed."[26] Thistlethwaite estimated the water depth during normal conditions to be 1.8 m to 2.4 m (6 feet to 8 feet).

Active forest fires were visible from the survey camps at all times during the entire 1950 boundary survey. Thistlethwaite reported being surrounded by four large fires for an extended period of time.[27] This caused considerable anxiety for the crew as they ran ahead of and into fires. Some of the boundary line they cut traversed through recently burned areas, while other sections of the line burned after the survey crew moved north. Thistlethwaite and his crew threaded a burning needle and unknowingly survived one of Canada's worst firestorms. The largest, continuous recent burned area that the survey crew walked through along the boundary occurred in Townships 99 and 100.[28] This was the Chinchaga River Fire (Fire 19). Interestingly, Thistlethwaite reported observing burns of different ages everywhere. It is quite probable the Chinchaga River Fire burned over areas previously burned.

Thistlethwaite made arrangements to be picked up by floatplane on Ekwan Lake on September 23. After crossing the Hay River, the survey crew travelled west and arrived at Ekwan Lake in the afternoon of September 22. According to Thistlethwaite, a very large and fast-moving fire chased his survey crew during their three-day trek to Ekwan Lake. "It became apparent later that the party actually had travelled just under the nose, as it were, of the fire and that the trail was burned out a few days after the former's passage. Smoke conditions were severe throughout the trip," he reported.[29] The *Edmonton Journal* described this large fire as one of the worst fires burning in British Columbia.[30]

The plane did not arrive at Ekwan Lake as planned. It was too smoky. A floatplane finally landed after being led into the lake by radio communication seven days later, and again the following day on October 1. The crew, personal gear, and equipment were flown to Charley Lake just northwest of Fort St. John. One of the guides led the horses by trail from Ekwan Lake to Beatton River, where they were trucked back to Fort St. John.

Each time Thistlethwaite's survey crew crossed a river, he documented observational evidence of the regional drought. Thistlethwaite's observations confirm that the Chinchaga River Fire was not the only large fire burning in western Canada's boreal forest in 1950. Fires roamed the parched landscape around them, but it was not until the big wind blew for three days that the firestorm erupted.

Large, catastrophic fires resist control because of winds. It is always the wind. Fuel can be managed, weather cannot. LaFoy had few options when the big wind arrived. He chose to stay and defend. Without defensive prevention strategies already in place, LaFoy was challenged to protect his community. Wind brought fire to LaFoy's district; when it stopped, the fire too stopped.

SEVEN

POLICY CHANGES

Policy changes predictably follow bad fire seasons. Such was the case in 1950. With a reported 314 fires and 591,264 ha burned, the 1949 fire season was one of the worst fire seasons experienced by the Alberta Department of Lands and Forests.[1] Settlers caused approximately 24% of these fires and 42% of the total area burned.[2] This was a good statistic for settlers interested in using fire to remove trees to plant crops, but not a welcomed statistic for Eric Huestis.

On February 28, 1948, Ted Blefgen, the director of forestry suffered a heart attack that forced him to resign. This provided a timely opportunity for Huestis to complete his rise through the ranks of the Forest Service. In March 1949, the Alberta Legislature passed a new act called the Forests Act to replace the Fires Act.[3] The government also divided the Department of Lands and Mines on April 1, 1949 into two separate departments: Mines and Minerals, and Lands and Forests.[4] Empowered with new legislation and a new dedicated department, Huestis, then acting director of forestry, commanded his troops to enforce the new act during the spring of 1949. The results were astounding. From April 1, 1949 to March 31, 1950, fire guardians (persons designated to enforce the Forests Act and regulations) and the Royal Canadian Mounted Police made 189 prosecutions under the new Forests Act.[5] The resulting 167 convictions represented 41% of the total number of forestry- and fire-related convictions in Canada in 1949.[6] Most of the convictions were against settlers who had started land-clearing fires without permits. Sixty-five of the prosecutions in Alberta were made in the Northern Alberta Forestry District (NAFD) by the Forest Service, and they resulted in 56 convictions.[7] The Forest Service also obtained an additional 60 convictions under the Game Act.[8] Huestis detailed the success of his enforcement campaign in the *Annual Report of the Department of Lands and Forests* for the fiscal year ending March 31, 1950.[9]

Despite the increased enforcement, the 1949 fire season was disastrous. In their annual report of forest fire losses in Canada, the federal Forestry Branch described the difficult spring fire season in Alberta: "Unfavourable moisture conditions at the time of the freeze-up in 1948 were followed by a spring of exceptional drought and warmth, and of high winds. This disastrous combination resulted, during the months of April and May, in nearly one million acres—a large proportion of the season's total loss."[10]

Large areas of merchantable timber valued at four million dollars (in 1950 dollars) burned during the 1949 fire season. J.L. Jansen, Alberta's chief timber inspector, did not want this to happen again. The protection of merchantable timber resources therefore became a fire suppression priority in 1950. This policy change and corresponding success in minimizing the merchantable timber losses for 1950 to one million dollars came with a cost—the loss of more structures and personal property. Because the Alberta Forest Service considered the Wanham, Newbrook, and Fishing Lake Fires as settlement fires, they provided little assistance to help the settlers. The trade-off of protecting values-at-risk still challenges fire managers today. The decision-making process becomes particularly difficult when fire threatens important non-timber values.

Armed with new regulations for the prevention of forest and prairie fires, established by Order-in-Council on May 15, 1950, Huestis seized the opportunity to change the Forest Service more and launch his legacy as the new director of forestry. He initiated a new policy during the spring of 1950: burning permits could only be issued by Alberta Forest Service personnel.[11] More importantly, before issuing a permit to burn slash or debris, the forest ranger had to first determine whether it was safe based on a site inspection of the hazard conditions. This policy change reaped immediate benefits. During the 1950 fire season, settlers accounted for approximately 15% of the fires, and only 2.5% of the area burned in Alberta.[12]

The new regulations specified no person could start a fire during the fire season (April 1 to November 13) to clear land for agricultural purposes unless he first obtained a permit issued by a fire guardian, who usually was the district forest ranger.[13] The permit required settlers to follow specific conditions before and during the fire. The area to be burned had to be completely surrounded by "a guard consisting of snow or water, or ploughed land not less than thirty feet in width and free of inflammable matter or waste."[14] Not surprisingly, settlers preferred to establish fire-guards using a plough. During the burning operation, a minimum of three adults equipped with two round mouthed shovels, one filled water barrel (45 gallons), an adequate supply of water, and four sacks were required to guard the fire.[15] These requirements looked good on paper, but were seldom fulfilled.

To provide support for this new system of issuing burning permits, an aggressive fire prevention campaign began in the spring of 1950. Fire prevention radio announcements and newspaper advertisements occurred throughout the province. One such newspaper article, titled "Forest Fires Are Costly," appeared in the *Battle River Herald* on July 27, 1950.[16] It stated the usual message: many fires are human-caused and preventable, and, in 1949 alone, 7,082 fires destroyed 2,728,656 acres (1,104,248 ha) of forests in Canada. This equated to half of what the pulp and paper industry consumed in the same year. The writer questioned why the human-caused fire problem persisted despite an increased suppression capability. Both the average fire size and the total area burned were about half of that reported in 1918; however, the number of reported fires each year remained relatively constant over the last thirty years. Preventing human-caused forest fires, in particular those caused by smokers, campers, and settlers, was the obvious solution, since only 19% of the fires reported were lightning-caused. Education would be the tool to deliver this message.

On September 21, 1950, the *Peace River Record-Gazette* reported that the Canadian Forestry Association, in co-operation with the Alberta Forest Service, completed their annual educational tour of the Peace River Country.[17] This "Caravan Tour" encouraged the public to co-operate in the prevention of forest fires. It was a huge success. The educational material included short coloured movies, which were a big hit, especially in the northern communities. In Fort Vermilion alone, 325 people filled the community hall to watch the movies. One movie called *Dead Out* showed graphic images of the destruction from fire caused by cigarette butts, matches, campfires, and the burning of brush piles.

Huestis also served notice that the new program for issuing burning permits would be enforced vigorously. Forest rangers did not share the same vigour to enforce the law. In the Prairie provinces, they preferred to use an educational approach to prevention.[18] Personally talking to individuals and groups was more effective, the forest rangers felt, than enforcing the law. It also made living in the same community with these people a little easier.

Nevertheless, with or without permits to burn, settler-caused fires continued to escape and cause problems for the Forest Service. In 1948, only eight fire prosecutions were made in the NAFD.[19] The same number of prosecutions was made during the previous year. During Huestis's directorship, prosecutions under the new Forests Act increased to 189 in 1949,[20] and 194 in 1950.[21] Settlers complained about the pushback from the Forest Service. Politicians too started to feel the heat of discontentment from settlers who worked hard to forge a living by clearing land and growing wheat.

An article in the August 17, 1950 edition of the *Peace River Record-Gazette* included a reference to the severe enforcement of the regulations and the subsequent complaints from settlers. The article, "Nature and

People Combined to Cut Forest Fire Losses," suggested that the settlers should do more to prevent fires: "So far this year Alberta has been fortunate in the low state of its forest fire losses, a condition which is probably attributable to moisture supplies and human effort. This is a situation which is most encouraging and indicates there is hope the educational efforts of the Provincial Department are beginning to pay dividends. Regulations which have been severe and in some instances cause for complaint, yet if it can be proven these have partially resulted in the existing situation then our settlers may feel more inclined to cooperate."[22]

During their 1950 annual meeting, the National Research Council's Subcommittee on Forest Fire Research appointed a panel to make recommendations to improve wildfire prevention in Canada. The panel decided that they first needed to understand how fires were started and the relative importance of these causal agents.

The panel distributed a questionnaire consisting of five questions to forest protection agencies across Canada. The results from their 315 replies were published the following year in the Forestry Chronicle. Settlers, fishermen, road travellers, and berry pickers accounted for two-thirds of all human-caused fires in Canada. In the Prairie provinces, settlers started approximately 25% of all human-caused fires. Eighty per cent of these fires resulted from the burning of hay meadows and the use of fire to clear land (broadcast burning and predominantly the burning of debris piles and windrows).[23]

During some years, however, up to 25% of the reported fires had no known cause. In the Annual Report of the Department of Lands and Mines for the fiscal year ending March 31, 1943, T.F. Blefgen, director of forestry, reported his suspicion of a new incendiary agent causing fires in Alberta.[24] He believed pilots and passengers were discarding lit cigarette and cigar butts from planes flying between Edmonton and Fort St. John. In 1942, nine fires occurred in Alberta in a remote area where no lightning storms occurred. These fires were aligned with the radio beam corridor used by pilots en route to Yukon and Alaska. Blefgen had good reason to implicate cigar-smoking pilots flying above Alberta's boreal forest. During drop tests in the United States lit cigars survived drops from as high as 1,829 m (6,000 feet).[25]

The commercial companies flying north through Alberta reassured Blefgen they were not the culprit as they posted warning signs and provided disposal trays in their planes for cigarettes and cigars. Passengers flying in the Douglas built DC-3 aircraft, which began scheduled passenger service in September 1936, were unable to open their windows to discard cigarette or cigar butts, but the pilots could, and did. They often smoked and opened their cockpit window during flight. After discounting passengers as the problem, Blefgen surmised it was the pilots, in particular, the military personnel travelling to bases along the northwest air route, who were causing these suspicious fires. When Blefgen talked to the military officers, they promised to stop starting fires. Did discarded cigars contribute to shape the boreal landscape by starting fires in remote areas? Dominion forester Harry Holman believed they did. He concluded that many of the large, remote wildfires in the Northwest Territories in the early 1940s were caused by discarded cigar butts from aircraft.[26] The later development of pressurized cabins ended this practice.

Aboriginal people reportedly caused most of the human-caused fires in about one-quarter of the forest districts in Alberta.[27] They traditionally used fire to increase blueberry production, improve the quality of grazing lands, and remove vegetation to help them find Seneca root. Aboriginal people used the aromatic root of the Seneca snakeroot plant for medicinal purposes. This perennial plant is a member of the milkwort family. The genus *Polygala* refers to the milky secretions characteristic of many species of this genus. A tea made by boiling the Seneca root was used to treat a wide variety of ailments and diseases such as coughs, sore throat, colds, pneumonia, bronchitis, asthma, typhus, rattlesnake and insect bites, and rheumatism. Canada was a major exporter of Seneca root during the 1950s.

Some fire use applications revealed remarkable cleverness and creativity. Prospectors, for example, used fire to burn surface fuels and expose mineral veins. Using a new tool, such as prescribed burning, was considered an advantage in the competitive search for precious metals, particularly if you didn't get caught and fined. Trappers too used fire to remove marsh vegetation to help locate muskrat lodges.

Some secondary benefits of fire still surprise. On July 3, 2006, the 11,000 ha Hourglass Fire approached the community of Tumbler Ridge, located in the Foothills of the Rocky Mountains in northeast British Columbia.[28] Not all of the 3,500 residents evacuated from this once-thriving coal mining town were upset. The Dinosaur Discovery Centre in Tumbler Ridge showcased British Columbia's only collection of dinosaur bones. For paleontologists excavating these 97-million-year-old bones near Tumbler Ridge, the Hourglass Fire made their job a little easier by removing vegetation and organic material and exposing new evidence.

Using fire to clear land was an acceptable and cheap land-use practice in 1950. Settlers perceived forested areas as obstacles to agricultural expansion. They called these areas "bush," meant to signify a low-value resource best replaced with crops that put food on the table. The general public, understandably, shared this attitude. Despite regulations in effect to protect the forest from fire, the government encouraged settlement and land clearing by subsidizing wheat growers. The Canadian government guaranteed that the price of wheat would not fall below one dollar per bushel from August 1, 1945 to July 31, 1950.[29] To help protect livestock, bounties were placed on cougars and wolves. A pair of cougar or wolf ears fetched a bounty of fifteen dollars.[30] The value of a wolf fur was only four dollars. Thus began the collision of policies between agriculture and forestry. Wheat versus trees. White Zone versus Green Zone.

On January 29, 1948, an Order-in-Council was passed to reserve the timber-producing areas in the province and prohibit settlement within these areas.[31] This administrative land-use zone, which later became known as the Green Zone, allowed the Forest Service to focus fire prevention and timber management activities in these areas. The main agricultural and settled areas in southeast Alberta became the White Zone. Crown land planned for eventual settlement was classified as the Yellow Zone (Figure 7.1). Most of the Yellow Zone was later reclassified as the White Zone. Establishing a barrier between forested areas and settlement areas was seen as a solution to reduce fire losses from settlers using fire to clear land. Similar forest protection challenges occur today, except these frontiers of settlement are now called the "wildland–urban interface." This is where communities, homes, and other values interface and intermingle with flammable vegetation.

One of Canada's worst wildland–urban interface fires occurred on August 16, 2003 in Kelowna, when lightning started a fire in Okanagan

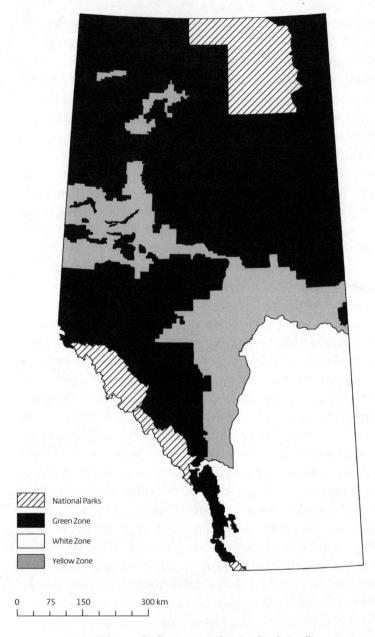

0   75   150   300 km

FIGURE 7.1   Green, White, and Yellow Zones, and National Parks in Alberta, 1948.

Mountain Provincial Park. Without any barriers to impede its spread, the fire raced into Kelowna, burned 239 homes, and forced the evacuation of 45,000 residents.[32] A record 2,500 fires during the 2003 fire season resulted in an unprecedented number of interface fires in British Columbia.[33] A review team chaired by the Honourable Gary Filmon completed a report titled *Firestorm 2003 Provincial Review*.[34] The review team's recommendations included the improvement of prevention by implementing fuel management in the wildland–urban interface.

The record set in 2003 would only last until 2009 when forest fires again threatened communities throughout British Columbia, including the city of Kelowna and homes along Okanagan Lake. Ash and evacuation orders again fell on the city. Resident Gordon Miller described the devastating return of fire to Kelowna: "It's kind of a nightmare to go through all this again."[35]

An increasing fire suppression technology, cooler and wetter weather conditions, and a lessened political will in the 1950s and 1960s negated the need to aggressively enforce the forest protection regulations. The focus remained on preventing fires through education and detection and initially attacking fires when they are small in size. All human-caused fires were considered a prevention failure. When Robson Black, general manager of the Canadian Forestry Association, returned from a visit to the Scandinavian countries in 1948, he made a convincing case to Alberta's forestry officials that education was the key to success for fire prevention.[36] His presentation made a lasting impression on Huestis, who in his first year as director of forestry, was convinced the Swedish model could be applied in Alberta. The model in Sweden was simple: educate all citizens, including children, about the importance of forestry to sustain the country's wealth and the subsequent need to be very careful with fire. Sweden, however, had more than education working to eliminate fire. Fewer lightning starts, weather less conducive to supporting large intense fires, more road access, and more managed forests helped to suppress fires. Nevertheless, if Sweden could eliminate human-caused fires from their forests, why couldn't Alberta?

This "can do" style was characteristic of Huestis, and it resonated throughout the Alberta Forest Service. In fact, later, the unofficial motto of the organization became "The Alberta Forest Service gets things done!"

Huestis used whatever means were available to him to advance the forest protection program. His determination to succeed matched his fierce loyalty to the government and, in particular, to Ernest Manning, who served as the premier of Alberta from 1943 to 1968. It did not matter to Huestis whether forestry accounted for almost half of Sweden's exports during the 1950s. If Sweden could do it, so could Alberta. But Alberta's economy, in comparison, was led by agriculture in the 1950s. While the oil and gas industry has fuelled Alberta's strong economic growth since then, forestry remains important. As Alberta's third largest industry, it provides economic support for 50 communities.[37]

As the Swedes continue to benefit from their success of eliminating the impact of human-caused fires on their forests, human-caused fires continue to challenge fire managers in Alberta.[38] The Swedes also continue to benefit from their success of eliminating the impact of human-caused fires on their forests. In Alberta, human-caused fires continue to challenge fire managers.

As the new director of forestry, Huestis proudly proclaimed the success of his burning permit policy change in the department's annual report for the fiscal year ending March 31, 1951.[39] The 1950 fire season reported 66 fewer fires and, more importantly, an area burned reduced by about half, compared to the 1949 fire season.[40] The year-end fire statistics were indeed impressive, except they excluded the Chinchaga River Fire and other fires during the 1950 firestorm. How could the largest fire in Alberta's history not even deserve a mention in the director's annual report? Huestis was aware of this fire because LaFoy pleaded with him to approve the hiring of a fire crew before the fire reached Keg River. Huestis could only proclaim success if the Chinchaga River Fire remained a ghost fire. Include it in the fire statistics, and a very different story would have unfolded. The 1950 fire season experienced fewer fires but actually burned more area than the disastrous 1949 fire season.

Huestis's decision to exclude the Chinchaga River Fire was not unusual. Non-actioned fires outside the firefighting protection area were not included in the annual report statistics. But the Chinchaga River Fire was not a typical ghost fire. Unknown to the Forest Service staff, the fire did burn into the firefighting protection area when it first crossed the provincial boundary.

Huestis learned from his predecessor, Ted Blefgen. In his 1944–45 annual report, Blefgen reported the unfortunate death of Forest Ranger D.B. Harrington from a stroke, but he failed to mention the two fire-fighter fatalities during the 1944 fire season.[41] Bad news seldom found its way into the annual reports.

The 1950 annual report for the British Columbia Department of Lands and Forests painted a different scene of the fire season in the Peace River Country. It described the 1950 fire season as "fortunate," except for the Peace River Section, east of the Rocky Mountains in the Fort George Forest District located in northeastern British Columbia.[42] Despite periods of high hazard throughout most of British Columbia, timely rains arrived and during the critically dry month of September, only a few of the usual lightning storms occurred. Little rain fell in the Peace River Country. This area remained dry and subsequently experienced a disastrous fire season. East of the Rocky Mountains in the Fort George Forest District, the fire season was described as severe. Approximately 94% of the reported 343,273 ha that burned in British Columbia in 1950 occurred in the Peace River Country in the Fort George Forest District.[43] Three fires averaged over 80,937 ha in size, and 15 fires accounted for 92% of the total area burned in the province for that year.[44] These fires burned predominantly in "immature timber" and "scrub-timber."[45]

Jerry McKee, deputy minister of forests for British Columbia, blamed homesteaders for starting three or more fires that merged to form the Chinchaga Firestorm. While addressing delegates at the 55th Pacific Logging Congress in Portland, Oregon, in 1965, McKee stated, "They succeeded that year in burning off 1,254,000 acres [507,474 ha]. They crossed the 120th meridian into Alberta and burned 2,500,000 acres [1,011,714 ha] in Alberta, and they crossed here on a 57 mile front. The smoke was so dandy that down in Chicago in the middle of the afternoon at the baseball game they had to turn on the lights. That is how good we are at making smoke up in this neck of the woods."[46] McKee was correct about the baseball game but it wasn't at Chicago. At 1400h on September 26, 1950, the lights were turned on at the Municipal Stadium in Cleveland, Ohio. Despite the unexpected darkness, the Cleveland Indians defeated the Chicago Cubs 2 to 0.

Although McKee's claim that several homesteader fires started the Chinchaga River Fire contradicts LaFoy's story of how the fire started, homesteaders did cause most of the fires burning in the region. Of the 1,515 reported fires in 1950 in British Columbia, 315 (approximately 21%) were assessed as caused by lightning. The majority of the 50 or more fires burning in northeastern British Columbia and northwestern Alberta in early June were human-caused fires (Table 1.1). Knowing these fires were all preventable frustrated the British Columbia Forest Service. Their disappointment is evident in the Department of Lands and Forests 1950 annual report. The chief forester stated, "The two major causes of Forest Service fire-fighting costs are campers and smokers, 43 per cent, and lightning, 33 per cent. The former is up 15 per cent over last year, and it is a sad commentary that carelessness on their part caused a needless fire-fighting expenditure of $63,780."[47] The higher suppression cost associated with the lightning-caused fires was due to the often remote location of these fires.

The chief forester also noted in his department's 1950 annual report that a period of high hazard occurred in the Fort George Forest District from June 8 to July 7 and during the month of September.[48] East of the Rocky Mountains the hazard was even worse.

The majority of the fires occurred in permafrost areas with no commercial forest value. Permafrost is a frozen layer in the soil that persists throughout the year. A downward trend in the fire danger began in early July and continued until September 9. Strong winds, low relative humidity, and a lack of rain quickly increased the fire hazard for the remainder of September. The report did mention when the major fire activity occurred: "New fires sprang up, old dormant fires revived, and settlers' burning-fires got out of control, particularly between September 19th to 25th, when numerous acreages were burned. These fires burning in the outlying areas, mostly in scrub-timber types, plus similar fires in Alberta, caused a widespread smoke haze which received considerable publicity."[49]

The Alberta firefighting policy allowed the Chinchaga Firestorm to continue to burn once it crossed the British Columbia–Alberta border and entered the Chinchaga River Valley in Alberta. Because fires in the NAFD located more than 16 km from the Mackenzie Highway were not included in the annual report statistics, the Alberta fire history database for the period prior to 1953 is incomplete. Forest protection agencies across Canada began recording forest fire statistics as early as 1918. As a result of the disastrous fire season in Canada in 1929, the Dominion Forest Service decided to standardize the compilation and reporting of these statistics for inclusion in their annual reports. However, many fires before 1960 were not reported. For example, the British Columbia fire history atlas missed several large fires that occurred in 1950 in the northeastern corner of the province. These fires are evident on the provincial forest cover map published in 1957.

In 1940, the Alberta government approved a policy decision to not fight fires in the northern part of the province.[50] The let-burn zone affected by the policy included the following areas: north of the north boundary of Township 92 between the 4th Meridian and Range 18, west of the 4th Meridian; north of Township 84 between Range 19, west of the 4th Meridian and Range 11 west of the 5th Meridian; and north of Township 92 between range 12, west of the 5th Meridian and the British Columbia–Alberta border. This area- or zone-based policy included a rule exemption allowing expenditures to fight fires that endangered lives or threatened property.

Eric Huestis, acting assistant director of forestry, sent a memorandum to Deputy Minister J. Harvie on March 29, 1945 requesting a northern extension of the southern boundary from Township 92 to Township 94 on the east and west sides of the let-burn zone.[51] The protection of timber resources and new settlements necessitated the policy change, which Harvie approved for the start of the 1945 fire season.

Further policy changes were recommended by Frank Neilson, the chief timber inspector for Alberta, in his September 27, 1945 memorandum to Eric Huestis. Neilson described the northern let-burn zone: "Over the greater part of this vast area few stands of merchantable timber remain. Over all of it, however, there is some type of cover made up mostly of Poplar, Willow, Alder and Black Spruce. Much of this cover is very sparse by reason of recurring fires. The great percentage of these fires is caused by the heedlessness of the native population and of trappers, both white and metis."[52] Neilson listed three main causes of these fires. The prescribed burning of meadows in the spring resulted in escaped fires burning large forested areas. The banks along sloughs were also burned to expose muskrat houses. These fires would smolder and hide in the peat and later resurface with the arrival of strong, dry winds. Lastly, abandoned campfires too often escaped to become large fires.

Neilson knew the department's suppression resources and budget severely constrained the policy options in the north. Nevertheless, he proposed a policy change that was adopted and became known as the 10-mile firefighting limit. The new policy required additional expenditures but Neilson believed the benefits outweighed the added costs. He recommended the conversion of part-time game guardians to full-time rangers so they could devote more time to protect the forest from fire. These rangers, he felt, should also be authorized to make expenditures to fight fires "in or directly threatening valuable stands of merchantable timber within ten miles of the Peace and Athabasca Rivers and within ten miles of the proposed Grimshaw highway."[53] Neilson's recommendation to convert part-time game guardians to full-time rangers was implemented in 1947.

In 1959, Frank Platt, assistant senior superintendent of the Forest Protection Branch in the Department of Lands and Forests, gave a public presentation in Edmonton at the Government of Alberta's first Natural Resources Conference. The printed version of Platt's presentation includes a reference to the 10-mile firefighting limit in northern Alberta in 1949. He stated, "The policy on fire suppression in the northern areas, i.e., north of Township 88, restricted suppression action to fires within ten miles of rivers or highways, except where lives or property were threatened."[54] Platt's recollection of the fire control policy in northern Alberta only refers to Township 88, yet the recorded official policy stated otherwise. No official policy could be found to indicate the northern boundary of the firefighting area moved further north during the 1945 to 1952 period. The historical records suggest a more plausible explanation. It was not always clear where and when to fight fire in the north. The requirement, though, to obtain approval from Edmonton for expenditures to fight northern fires was very clear. The ability to set provincial priorities and decide which fires should get resources is essentially the same approach used in 2014.

During a presentation to new forest officers at the first orientation course held in 1958 at the Forest Technology School in Hinton, Alberta,

Huestis provided an historical perspective and rationale for the 10-mile firefighting limit:

The government at the time was pretty short of money, and so they refused to let us spend any money fighting fires north of Slave Lake. Twenty miles north of Slave Lake, that was the cutoff, and 20 miles north of Peace River unless you could reach it by river, and we used rivers for navigation like the Peace and the Athabasca. If it was within 10 miles of the river you were allowed to go and fight it, but we didn't have much to fight it with so nothing was done. We lost huge areas of that north country purely and simply because they wouldn't let us touch it.[55]

Huestis was right. Many person- and lightning-caused fires roamed freely across the fire-prone forest landscape of northern Alberta. The policy change to fight all fires in 1953 was a timely political decision. The ongoing forest inventory program revealed large areas of merchant-able forest resources in the NAFD that warranted protection from fire. The forest industry, oil and gas industry, and northern residents were also becoming increasingly concerned about fires and the need to protect their investments. Although the policy change seemed reasonable in light of changing attitudes, implementing this policy was operationally not possible. The Forest Service lacked the resources and staff to success-fully fight all fires. This soon changed.

The 1950 fire season was not an anomaly. Extensive fires occurred not only in the previous year but again in 1953 from north to south. The 1949 and 1950 fires forced the Forest Service to reconsider its policy of allowing fires to burn when they were more than 16 km from a major transportation corridor or settlement. On September 16, 1952, the 16 km firefighting limit policy was dropped.[56] Forest rangers received orders to fight all fires regardless of their location during the next fire season.

In 1953, the Rocky Mountain Section of the Canadian Institute of Forestry submitted a review of Alberta's fire control program to Ivan Casey, the minister of the Department of Lands and Forests.[57] Known as The Fire Brief, this scathing report included 22 recommendations to improve the forest protection program in Alberta. The Forest Service was aware of the concerns of the growing forest industry and the Canadian Institute of Forestry's plans to submit their report. This was a contributing factor in the government's decision to drop the 16 km firefighting limit. *The Fire Brief* resulted in substantial increases in funding and support for fire control in Alberta. The report, though critical, arrived at a time when the government could afford to increase expenditures to strengthen the forest protection capacity and capability because of increased revenues from a booming oil and gas industry. Norman Willmore, newly appointed minister of the Department of Lands and Forests, also helped. From 1955 to 1963, he provided strong support and leadership to implement innovative improvements to the forest protection program. Willmore secured the required funds for more roads, lookout towers, firefighting and camp equipment, aerial patrols, ranger stations, and a new forestry training school at Hinton. These enhancements strengthened the Forest Service's capability to prevent, detect, and suppress fires.

According to LaFoy, he did submit a fire report on the Chinchaga River Fire to the Peace River Office.[58] Fire reports exist for the other 1950 fires in the Peace River District, but the Chinchaga River Fire report disappeared. LaFoy drew the boundaries of the fire on a map based on information he received from trappers who checked their traplines after the fire. He gave this map to Timber Inspector Larry Gauthier in Peace River.[59] This map, too, vanished.

Forest Ranger Mitchell decided not to fight the Whisp Fire after consulting with the fire inspector and the district forester. They felt it would be hopeless to prevent the logged area from eventually burning with the current rate of settlement. They also concluded the probability of containment was very low even if they could bring in firefighting resources from Fort St. John. Since the fire occurred outside the firefighting zone, the decision was made to let it burn freely.

Mitchell's report for the Whisp Fire included handwritten recommendations from the district forester: "This fire lies in Compt [Compartment] C of pre-organization chart and our policy approved by your office—was to fight fires here only if there was a reasonable chance of saving immature timber. The decision in this regard is shown in separate history. It is unfortunate that September was so dry and that

Mr. Hack was burned out. We can expect repetitions of fires such as this unless our policy changes and a major firefighting course is based in the Peace."[60]

After drilling 133 dry holes, geologists convinced Imperial Oil to drill one more well in 1947, which they nicknamed "Last Chance."[61] They drilled 1.6 km into rock formations they new little about, and found oil. This discovery set off a province-wide search for oil. The oil industry subsequently lobbied the Government of Alberta to allocate funds for base mapping to support increasing oil exploration activities. The government realized a provincial aerial photography and mapping program would also benefit the growing forest industry. In 1949, the Photographic Survey Corporation of Toronto began aerial surveys to complete base maps for the entire province and forest inventory maps for the forested areas south of the 57th parallel and north of the East Slopes Forest Reserves (about the 53rd parallel).[62] The base maps were produced using 1:40,000 scale aerial photographs. The forest inventory was later extended north of the 57th parallel, to cover forested areas of interest in northern Alberta.

Erling Winquest worked as an interpreter for the Technical Division in the Alberta Forest Service during the 1950s. Using a pocket stereoscope with two times magnification, he identified forest stands directly onto the 1:15,840 scale aerial photographs. The forest classification was then transferred from the aerial photographs to the 1:40,000 scale base maps using an instrument called a transfer scope.

With the first phase of Alberta's forest inventory completed in 1956, forest cover maps became available for the entire province except the East Slopes Forest Reserves and the national parks. The following year, the first provincial Forest Cover Map (1 inch to 16 miles) was completed for the province using the original forest classifications.[63]

The 1957 map includes six forest cover classes and two burn classes: coniferous stands up to 60 feet in height, coniferous stands over 60 feet in height, mixedwood stands up to 60 feet in height, mixedwood stands over 60 feet in height, deciduous stands up to 60 feet in height, deciduous stands over 60 feet in height, burns 1941 to 1957 inclusive, old burn and brushland. There are also five non-forest cover classes: agricultural and other improved lands, muskeg and marsh, rock barren, hay meadows, and barren above timberline. The large area classified as burns and old burn corroborates Neilson's description of the boreal forest and the awesome power of fire to sculpt and change this landscape (Figure 7.2). Practically all of the boreal forest in Alberta originated from fire. In natural evolving fire-dependent ecosystems, there is little chance of forests growing old. Fire chases and eventually finds the fuel that is ready to burn.

The Alberta Department of Lands and Mines published a map out-lining areas devastated by forest fires from 1938 to 1949 (Figure 7.3).[64] Areas burned under 405 ha in size are not included on the map. Large fires occurred within the Grande Prairie, Peace River, Swan Hills, Lesser Slave Lake, and Lac La Biche areas, and along the edge of the settlement areas. The published date of 1946 is likely incorrect (it should be 1949) because the map delineates the area withdrawn from settlement.

The 1949 and 1957 maps tell an interesting story. Much of Alberta's landscape is a product of human-caused fires. Earlier reports provide further evidence. The federal Forestry Branch of the Department of the Interior completed a report in 1916 titled "Timber Conditions in the Smoky River Valley and Grande Prairie Country."[65] Since 1886, forest fires burned approximately 83% of the study area. Settlers encountered difficulty finding forest stands old enough to provide timber for build-ing construction. The author of the report, J.A. Doucet, felt there was a need for a special organization to fight forest fires and protect the timber.

The 1957 forest cover map represents the first view of Alberta's for-ests before effective wildfire suppression began in about 1960. Absent from this map is the Chinchaga Fire. The 1950 photography of the Chinchaga River area was taken just before the fire occurred. Asked why the Chinchaga River Fire was never mapped, Winquest indicated that no photographs were available. Areas burned by fires greater than 65 ha were supposed to be flown and photographed to update the forest cover inventory. This procedure is noted in the 1968 Department of Lands and Forests publication *Alberta Forest Inventory*.[66] Updated photography was never acquired because the Chinchaga River Fire remained a ghost fire until 1958, when Reg Loomis started asking questions about the report submitted for another fire named 28A-2.[67]

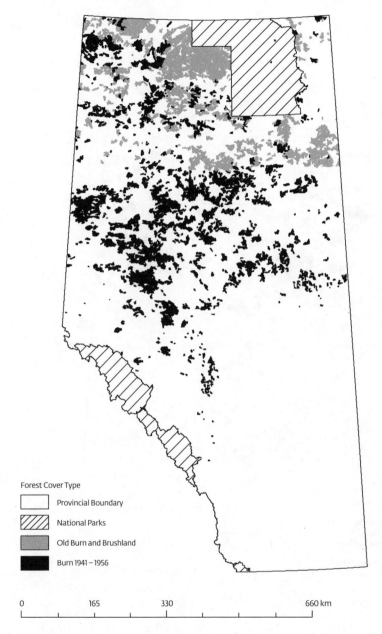

Forest Cover Type

☐ Provincial Boundary

▨ National Parks

▨ Old Burn and Brushland

■ Burn 1941–1956

| 0 | 165 | 330 | 660 km |

FIGURE 7.2　Burned areas as delineated on the 1957 forest cover map of Alberta.

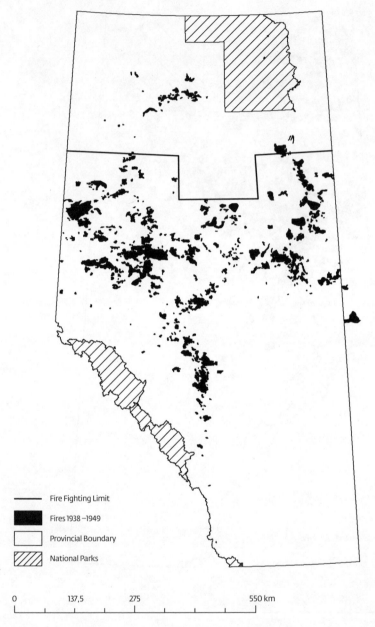

FIGURE 7.3   Known location of fires greater than 404 ha (1,000 acres), 1938–1949. Alberta Forest
Service, 1949.

In February 1951, Jack Grant replaced Frank LaFoy as the forest ranger stationed at Keg River.[68] Jack worked on Fire 28A-2, a difficult 60,605 ha, lightning-caused fire reported on July 31, 1958. After the fire, Jack was transferred to Edmonton to work as an aircraft dispatcher. Reg Loomis also worked for the Forest Service in Edmonton. Huestis hired him in the fall of 1949 as the province's first professional forester to oversee Alberta's first broad-scale forest inventory and implement sustained yield forestry practices. Loomis noted the report submitted by Grant for Fire 28A-2 referred to a large unknown fire to the north of the Chinchaga River. Loomis asked Grant what fire he was talking about. Grant indicated it was the big fire that burned from Rose Prairie, British Columbia to Keg River, Alberta. Loomis replied, "What fire?" "Well," said Grant, "it burned over 3 base lines."[69] The distance between two baselines is 38.8657 km. From the high points of land, Grant indicated he could see from one baseline to the next. Loomis was stunned. According to Grant, "this really blew his mind."[70]

Lou Babcock, assistant forest superintendent at Peace River, was the fire boss in charge of Fire 28A-2. Transporting supplies and 500 firefighters to this particular fire was, as he described it, one of the worst experiences of his firefighting career.[71] Firefighting efforts did little to slow this fire. And too often, the firefighters spent more time seeking shelter and running from the fire than fighting it.

Babcock was pleased when a Forest Service Beaver aircraft, CF-DJM, arrived at the Eureka River base on the evening of September 7 to help transport supplies and men to the fire. On the morning of September 8, pilot Maynard Bolam flew the single-engine plane with a load of aviation gas and regular gas to the base airstrip at the fire. He returned to the Eureka River base with a sick firefighter and his crew leader. At 1335h, Bolam departed the Eureka River base for another 20-minute supply run to the fire. On board was 1,000 pounds of flour, a barrel of naphtha fuel, and the crew leader returning to the fire.[72] CF-DJM never arrived at the base airstrip. They likely lost vision while flying through the smoke.

One of the most exhaustive air searches in Canada failed to find the plane. The crash site was not found until 19 years later when an Alberta Forest Service helicopter flying in the area spotted the wreckage. Since 1943, 22 wildland firefighters have died during wildfire suppression operations in Alberta, including the three young men who died while fighting fire—one in 1943 and two in 1944.[73] Fighting fire can be a dangerous job. It is fortunate no lives were lost during the 1950 Chinchaga Firestorm.

EIGHT

RING OF STEEL

Ed Olney was a pioneer in the forest industry in central Alberta. After the Second World War, he moved from Drayton Valley and established three sawmills and a planer mill in the Slave Lake area. In the spring of 1946, a fire near Snipe Lake, south of High Prairie, grew quickly out of control. The Alberta Forest Service conscripted men and equipment from Olney's company to help fight the fire. The fast-moving fire chased one of Olney's three bulldozers for 16 km before the operator and his assistant sought refuge in a clearing. Two days later, they walked out to the highway. When the hot ashes burned the soles of their boots, they cut strips from their jackets, tied them around their battered boots, and continued walking.[1]

The Snipe Lake Fire burned out of control for several days. Mason Wood, the Slave Lake forest superintendent, decided he urgently needed to map the fire and the location of the bulldozer guards. He asked Olney to fly over the area and take photographs. Since the department firefighting policy did not allow Forest Service staff to hire aircraft, Wood asked Olney to pay for the plane and add the cost to the bulldozer invoice. After a long day, the fire perimeter, fireguards, and spot fires were mapped. The creative billing scheme worked well until the Forest Service Headquarters in Edmonton received a copy of the map. Wood was immediately questioned how he obtained such an accurate map so quickly. He had no choice but to tell the truth. Impressed with the accuracy and speed of the map production, Ted Blefgen, the Forest Service director, agreed to pay for the use of the aircraft, despite the breach of policy.

After the Snipe Lake Fire, the Forest Service decided to use aircraft for the reconnaissance and mapping of large fires. This included manually sketching fire perimeters, bulldozer guards, and spot fires, and taking aerial photographs.

In 1964, the Forest Service again asked Olney for assistance, not to fight large fires but to help keep them small. They asked Olney to send his logging crew to Hinton to take a firefighting course. Olney agreed. Having watched a fire destroy his first sawmill in Drayton Valley, he knew the importance of maintaining a strong fire suppression capability. On October 1, 1960, the Department of Lands and Forests officially opened the Forestry Training School in Hinton.[2] In 1962, fire control training was extended to people outside the Forest Service. Olney's logging crew became one of the first certified initial attack crews capable of being deployed to a fire by helicopter. Olney described the helicopter as a "wonderful machine for fighting fire." "It also made a dandy fishing boat," he recalled.[3]

During the 1950s, bulldozers became an important tool to control and suppress fires. To survive, a fire requires fuel, heat, and oxygen. This is called the fire triangle. Remove any side of the triangle and fire will die. In Alberta's boreal forest, bulldozers are very effective at removing fuel when constructing fireguards. These fuel breaks vary in width from a single dozer blade to 150 m or more, and can be constructed close to or away from the fire, depending on the fire's intensity.[4] In dense stands of tall trees, two blade widths are required to prevent burning trees from falling across the guard and starting a fire on the other side. Wider is not necessarily better. The objective is to build the narrowest guard to control the fire. This reduces the cost, environmental damage, and, more importantly, time. When bulldozer operators construct a guard around a fire, the "ring of steel" is complete and the fire's status changes from "out of control" to "being held." This indicates there is no or little likelihood of the fire getting larger.

How well a guard contains a fire is largely dependent on the support resources following the bulldozers to prevent the fire from spotting across the guard. The fuel between the guard and the wildfire is often burned out in an attempt to strengthen the guard and starve the advancing fire of fuel.[5]

Bulldozer guards are constructed by blading the ground and removing all flammable vegetation and organic material. Only mineral soil is left behind. A team of three bulldozers can be very productive at building guard. The first bulldozer operator knocks down the trees, ensuring they do not fall into the fire. The second bulldozer operator pushes the trees away from the fire. Completing the guard to mineral soil is the responsibility of the third bulldozer operator. This operator also constructs landing pads for helicopters and digs seepage holes to collect water for firefighters with pumps.

Guards can be constructed by firefighting crews using hand tools such as shovels, rakes, pulaskis, and saws, but in the boreal forest, this becomes difficult in areas where deep organic soils predominate.

The flammable vegetation can be moved to either side of the guard, but is typically moved to the unburned side. There are some exceptions. For example, it may be desirable to move the flammable vegetation to the inside of the guard, to increase fire intensities, and ensure the success of a burn out operation. On steep slopes, it is impractical to push debris uphill, and, if it is still burning, the debris must be moved to the fire side of the guard.

The first bulldozer was a modified Holt farm tractor. In 1905, Richard Hornsby & Sons from Grantham, England, invented the "caterpillar" track.[6] This revolutionary invention changed not only the agriculture industry but also how wars and fires are fought. Hornsby sold his patents to Benjamin Holt in 1914 for eight thousand dollars.[7] In 1925, Holt's California-based company merged with the Best Gas Tractor Company to form the Caterpillar Tractor Company.[8] The eventual installation of a thick metal front plate converted the large and noisy tractor from a ploughing and hauling machine to a powerful earth-moving machine. During the postwar period, the construction of roads, canals, dams, pipelines, communication infrastructures, and a growing forestry industry created a voracious appetite to move earth. Each year, improvements were made to the large, tracked excavation machines. By 1940, owning a Caterpillar bulldozer was akin to owning a small bank.

Although bulldozers could build fireguards considerably faster than firefighters with hand tools, policy did not authorize their use in the Northern Alberta Forest District until 1951.[9] Nevertheless, bulldozers did find their way onto fires during the 1940s to help protect life and property. With firefighter wages averaging only twenty cents per hour, the Government of Alberta was reluctant to pay the higher cost of using bulldozers.[10] It became difficult, however, to find able men to fight fire during the Second World War. Policy dictated what should happen; budgets and politics dictated what really happened. Technology helped to drive this disconnect.

The Canadian Institute of Forestry Standing Committee on Forest Fire included agency reports in a special supplement to the March 1951 issue of the *Forestry Chronicle*. In their report, the Alberta Forest Service mentioned great success in the use of Caterpillar tractors with bulldozer blades during the 1951 fire season in the Northern Alberta Forest District.[11] Since then, the bulldozer has become the tool of choice to build fireguard in the boreal forest of Alberta.

In 1950, the priorities for allocating resources to fight fires were made on a fire-by-fire basis. The same approach is used today. During a multi-fire situation, high-priority fires receive the most suppression resources, while little or no suppression resources go to low priority fires. Protecting life and property is the first priority for fire management agencies. To protect life, the fire must be stopped before it overruns people, or the people in the path of the fire must be evacuated. People can be moved with relative ease and success if they all reside in one location and emergency response teams have sufficient time to execute the evacuation plan. The challenge becomes more daunting when a large, fast-moving fire threatens homes and subdivisions scattered throughout the forest. People who flee a fire often drive on the same smoke-shrouded road used by firefighters rushing to that fire. Unable to see because of the smoke, panicked drivers crash their vehicles and block the escape route for others. Hikers, backcountry campers, and other recreational users also need to be located and evacuated.

In Canada, the provinces of Yukon, Saskatchewan, Manitoba, Ontario, and Quebec, the Northwest Territories, and the national parks have extensive areas where little or no wildfire suppression occurs unless wildfires threaten remote communities or other high values-at-risk. The Canadian Interagency Forest Fire Centre (CIFFC) reports the area burned each day from full and modified response wildfires. Not surprisingly, the area burned in the limited action or modified suppression zones is usually larger than the area burned in the full suppression or intensive fire management zones. The modified suppression strategy allows wildfires to burn with minimal interference if values-at-risk are not threatened. Not all of these fires are reported to CIFFC. In 2003, 650,000 ha burned in northern Ontario and Manitoba, near Hudson Bay. These fires did not get mapped. The 1.65 million-hectare area burned in Canada, as reported by CIFFC for 2003, should be 2.3 million hectares.

Because fires are a natural process, it is not ecologically desirable to remove all fire from fire-dependent ecosystems. Nor is it economically feasible to eliminate all fires from the landscape.[12] The challenge lies in finding an appropriate balance. To do this, fire managers must make decisions that acknowledge the limits of suppression effectiveness. It may be unnecessary and inappropriate to dispatch endless suppression resources to all fires and at all times. During its major runs, the Chinchaga River Fire released energy equivalent to a Hiroshima bomb exploding every three minutes. Fighting a fire with this energy release rate is best done by stepping back, protecting the values-at-risk if possible, and waiting for the dragon to sleep.

Fire can be beneficial at the right time and place. It can also be a very powerful and destructive force at the wrong time and place. Economic and social considerations therefore tend to override ecological considerations. Sometimes, social considerations create catastrophic consequences later. Had a prescribed burn been implemented in Okanagan Mountain Provincial Park as planned, the impact of the lightning fire that started on August 16, 2003 may have been reduced.[13] Unfortunately, the public and municipal staff were not supportive of the use of prescribed burning to manage forest fuels. The proposed prescribed burn would have created a strategic fuel break for one of the most destructive interface wildfires in Canadian history. Lessons were learned. Accept a little smoke from a controlled prescribed burn for fuel management, or a lot of smoke and burned houses later from an uncontrolled wildfire.

The 2003 Kelowna Fire was not an isolated event. In 2011, a more devastating interface wildfire occurred in Alberta. Between 1980 and 2007, 547 evacuations due to wildfires occurred in Canada.[14] An average of 7,469 people were evacuated each year—a small number compared to the amount of fire activity across Canada.

Beginning around 1995, fire management agencies in North America shifted their focus to become structure-protection agencies. They did not choose to make this change. People moving to flammable forested areas assumed they were safe. They expected it. Unfortunately, protecting properties that have little or no chance of surviving a fire robs valuable suppression resources needed to contain the fire. Fire managers are also increasingly being asked whether the cost of suppression exceeds the value of the resources being protected. This return on investment question is valid, but the valuation of resources, in particular, the non-timber values, remains a challenge.

Despite the increasing expenditures to fight fire, the declining marginal returns on investment suggest fire management agencies in Canada and the United States are at a critical juncture. The area burned in Canada is predicted to be more than double by 2040 due to a threshold effect occurring when the future fire load overwhelms the capacity of fire management agencies.[15]

If climate change brings more fire starts and escaped fires, and larger and more intense firestorms, change is inevitable. We must coexist with fire by being better prepared when it arrives. In 2007, the Food and Agriculture Organization of the United Nations released a report in

Rome titled "Fire Management—Global Assessment 2006."[16] This report identified the need to enhance fire prevention by developing effective fuel management and prescribed burn programs to reduce the potential losses from wildfires. It also concluded today's fire management strategy and tactics cannot be sustained in the future because of the exorbitant costs required to attain the performance measures in a changing fire environment.

Dave Martell, at the University of Toronto Fire Management Systems Laboratory, described this moment of truth during a teleconference presentation to the fire management community in Canada on January 24, 2007.[17] The title of this national fire management conversation sponsored by the Canadian Interagency Forest Fire Centre, summarized the dilemma concisely: "Forest Fire Management in Canada—We Have to Stop Behaving Like This."[18] Martell believes fire management agencies need to gradually reduce the area where they practice fire exclusion while increasing the area where fire can play a more natural role. He challenged the notion that current fire management strategies are sustainable or fully compatible with the principles of sustainable forest management. This approach strives to maintain or enhance the long-term health and productivity of forest ecosystems to ensure a flow of ecological, social, and economic services into the future.

In the early 1950s, aircraft were not yet fully integrated in Alberta's fire suppression program. As a result, the Forest Service relied on firefighters, bulldozers, and burn out operations. Without the use of aircraft, communities threatened from fire were protected by constructing guards and setting backfires. This practice stopped in the 1960s when both funds and aircraft became readily available to fight fires. Less severe fire weather also occurred in the 1960s and 1970s.

In the boreal forest, backfire operations are used to indirectly attack the head of a fire when direct attack is not possible. Backfiring can be conducted along constructed guards or natural features such as lakes, rivers, or fuel type changes. It is impressive to watch a backfire operation. A large high-intensity fire creates its own weather. As the hot air rises, the surrounding air is pulled into the fire to feed its enormous thirst for cooler air. These in-drafts can be very strong. The power of the fire can literally draw in the backfire against the wind. If the power of the wind exceeds the power of the fire, the backfire operation fails. Sometimes a large backfire is ignited to create a separate large firestorm to, in effect, pull the advancing fire. This is done, for example, to strategically steer the head of the fire towards a river, lake, or ridge top. Backfires can also be used to slow the advancing fire and make the smoke column stand up, thereby increasing the visibility for air operations.

When the power of the wind exceeds the power of the fire, the fire is classified as a wind-driven fire. These fires are characterized by their bending and low-lying smoke columns. Wind-driven fires are difficult to suppress because of the dense smoke, reduced visibility, and high rates of spread. Most of the large spring fires in Alberta's boreal forest are wind-driven fires. When the power of the fire is greater than the power of the wind, the fire becomes convection driven. The vertical smoke columns above these fires create strong in-drafts and turbulence. Burning embers can be carried very high in the convection column and then transported up to 10 km ahead of the main fire front. Since visibility is relatively good for air operations, burn out and backfiring using aerial ignition can often be conducted on these fires.

It is common for sparks or embers to be carried by the wind up to 100 m ahead of the advancing fire front.[19] During high-intensity crown fires these embers can be transported up to 1,000 m.[20] The spot fires caused by these embers are usually overrun by the advancing fire front. Long-range spotting up to 10 km in distance, though infrequent, can result in new fires well ahead of the main fire front.[21]

Burn out operations were an important component of early fire prevention programs. In 1908, the forest ranger at Cypress Hills Forest Reserve in southeastern Alberta received instructions to plough a guard four furrows in width around the entire reserve.[22] A second guard was ploughed a distance of four rods (one rod = 5.0292 m) outside the first guard. During a calm day, and with the help of two or three men, the forest ranger would burn the grass between the two guards. In the Spruce Woods Forest Reserve in Manitoba, guards were ploughed along each side of the Canadian Northern Railway.[23]

Burning meadows was also a common practice. Meadows around forest reserves were burned in the spring when snow still lingered in the surrounding forest. The inspector of forest reserves for the Department of the Interior made an interesting observation in 1908 during his inspection of Turtle Mountain Forest Reserve in Manitoba: the roads ran north and south, in the same direction that most of the fires were spreading.[24] He subsequently advised the forest ranger to build a road to traverse the reserve from east to west. The inspector noted the potential use of the new road as a guard to set backfires from.

The terms *burn out* and *backfire* are sometimes used interchangeably, even by fire management staff. A burn out operation attempts to burn out fuel between a guard and the fire. The burn out fire generally spreads with the wind. A backfire also attempts to burn out fuel but spreads against the wind towards the wildfire. Backfires are usually set far away from an advancing fire. The terms *counter fire*, *back burn*, *burn out*, and *burn off* all add to the confusion. Nevertheless, all of these operations refer to the skillful and knowledgeable application of fire to fight fire.

Although fighting fire with fire can be very effective, it is usually applied as a suppression strategy of last resort. The risks and benefits of setting a backfire need to be weighed against the risks and benefits of not setting a backfire. There is no specific cookbook to provide instructions on when and how a backfire should be ignited for a particular fire and under what precise conditions it should occur. Each fire situation is different. Even though fire managers have considerable flexibility and discretionary authorization to use fire as a suppression tool, good judgement, common sense, and experience are paramount.

Sometimes backfiring fails. The public, too, sometimes fails to understand why firefighters intentionally set these fires in their effort to protect their life and property. The success of a backfire operation on August 6, 2000, on a fire in western Montana was contested by homeowners and an insurance company. Concerned that the Spade Fire in the Bitterroot Valley would spread north towards communities, the Incident Command Team decided to ignite a backfire. A group of landowners who lost property filed a claim in the District Court of Montana, stating that the backfire, not the main fire, burned their homes. Chief Judge Donald Molloy ruled in favour of the firefighters. He concluded the United States government was not negligent and responsible.[25]

Increasing expectations to protect life, property, and resource industry investments were the impetus for the Alberta government's decision in 1956 to purchase its own aircraft and to use seasonal and casual contract aircraft.[26] Three fires burned 24,686 ha in the North Western Pulp and Power Ltd. Forest Management Agreement area in 1956.[27] Signed in 1954, this was Alberta's first forest management agreement. The company became concerned when the smoke cleared from these early summer fires. Almost 3% of their licensed area burned before construction of their pulp mill in Hinton was even finished.[28]

The following year, the Forest Service purchased its first aircraft, a Helio Courier CF-IYZ.[29] In 1958, a contract fleet of Stearman aircraft from Harrington Air Services was dispatched to fight three fires in the Edson Forest.[30] PBY Canso (first used in 1966), Douglas B-25 (first used in 1967), and B-26 (first used in 1970) aerial tankers soon replaced the Stearman aircraft.[31] In 1967, the durable Canadair CL-215 amphibian air tanker became the only aircraft in the world specifically designed and built to fight fires.[32] The first production delivery was made in 1969 to France. This piston engine water bomber, referred to as the "yellow duck," took off and landed on airstrips or open water, and could scoop 5,455 litres of water into two tanks in about 10 seconds.[33]

In 1994, Canadair began delivering the faster and more powerful CL-415 water bomber.[34] Two Pratt & Whitney turbo prop engines increased the water scooping capacity of this aircraft to 6,137 litres. Other aircraft joined the elite global firefighting air force. The Douglas DC-6 and DC-7, Martin Mars, and Ilyushin IL-76 are all commercial aircraft converted into large air tankers to fight forest fires. With a 61 m wingspan, the Martin Mars became the largest production seaplane in the world. Howard Hughes's prototype H-4 known as the "Spruce Goose," and made famous in the movie *The Aviator*, was a larger seaplane, but flew only once, two years after the war.

In 1938, the United States Navy ordered a prototype Martin Mars for use as a patrol bomber seaplane. On November 5, 1941 the Glenn L. Martin Factory in Little River, Maryland, launched an experimental XPB2M-1 aircraft nicknamed "Old Lady."[35] On December 5, 1941, two days before the Imperial Japanese Navy attacked Pearl Harbor, one of the large wooden propeller blades on "Old Lady" flew off an engine during taxiing tests and tore into the fuselage, narrowly missing the flight engineer. During the six months it took to complete the repairs, the

United States Navy reconsidered the plane's military role as a result of the losses they incurred during the Pearl Harbor attack. The Japanese clearly demonstrated the effectiveness of using fast carrier planes to deliver bombs. The "Old Lady" in comparison was a slow and easy target for the enemy. The Navy decided to use her as a long-distance carrier to transport supplies and troops—an alternative to using merchant ships to cross the submarine-infested waters of the Atlantic Ocean. With gun turrets and bomb bay removed, the "Old Lady" was redesignated as a transport XPB2M-1R aircraft and assigned to haul cargo on the California to Hawaii route.[36]

In January 1945, the Navy ordered 20 improved Martin Mars seaplanes.[37] These new JRM-1 transport aircraft boasted a new single vertical stabilizer, larger fuel tanks, longer hull, larger cargo doors, two reinforced deck levels, and 18-cylinder Wright Cyclone R-3350-8 engines to provide increased takeoff weight. The first JRM-1, named the Hawaii Mars, was delivered July 21, 1945, about a month before the atomic bombing of Hiroshima and Nagasaki.[38] Four other JRM-1 aircraft were built and delivered to the United States Navy in 1947.[39] When Japan surrendered on August 14, 1945, the Navy cut their order for JRM-1 aircraft to six. Thus, only six aircraft left the production line in Little River. Each ship was given a name: the Hawaii, Philippine, Marianas, Marshall, a second Hawaii, and the Caroline. In May 1948, the Caroline Mars was the last Martin Mars aircraft and the only JRM-2 model delivered to the Navy.[40] The JRM-2 aircraft included Pratt & Whitney R-4360 Wasp engines, each producing 3,000 horsepower on takeoff.

The original Hawaii Mars crashed and sank during a landing attempt on Chesapeake Bay two weeks after its first flight on July 21, 1945.[41] A second Hawaii Mars, still flying in 2014, was subsequently built the same year. On May 5, 1950 the Marshall Mars caught fire near Hawaii. The crew managed to make an emergency landing at sea and exit before the fire destroyed the aircraft.[42]

In 1959, a group of forest companies in British Columbia formed Forest Industries Flying Tankers Inc. This consortium purchased the last four Martin Mars planes—the Hawaii, Philippine, Marianas, and the Caroline—and then converted them to water bombers.[43] On June 23, 1961, while fighting fires, the Marianas Mars struck trees and crashed.[44]

The following year, the Caroline Mars was destroyed by Typhoon Freda while parked on its beaching gear at the Victoria International Airport on Vancouver Island.[45]

After fighting forest fires in British Columbia for 40 years, Coulson Aircrane Ltd., a local Vancouver Island company purchased the remaining two giant flying boats from Forest Industries Flying Tankers Inc. in April 2007, with the intent to the continue the tradition of using them to fight fire.[46]

The Coulson flying tankers remain based on Sproat Lake near Port Alberni on Vancouver Island. These 200-feet wingspan seaplanes need water to land and take off. They can scoop 27,000 litres of water in about 30 seconds.[47] On October 24, 2007, the Hawaii Mars left Vancouver Island and flew to California to help fight fires burning in the San Diego area for four days.[48] The Hawaii Mars was delayed in flying firefighting missions because it struck two birds on route to San Diego. Repairs and an inspection were required before it could begin dropping water on the Harris and Witch Creek Fires.

The Russian Ilyushin-76 supertanker is the largest firefighting aircraft in the world. It can drop an amazing 42,000 litres of water or retardant on a fire.[49] Unfortunately, few air tanker bases in North America can accommodate an aircraft of this size and weight. As well, there is concern about its capability to operate safely in terrain requiring downhill drops.

In 2002, a California-based engineering company proposed the concept of using airships capable of carrying one million litres of water.[50] These buoyant airships would hover above the fire and release their payload as continuous curtains of rain. The plan looked good on paper, but this was not the first time engineers failed to grasp the realities of the power of fire. Second World War allied bomber pilots, returning from their missions, had difficulty landing in airfields in England because of fog. Several engineers proposed a simple solution to burn off the fog: build large fires at the end of the runway using natural gas jets. The first pilot to test the new system nearly crashed when he encountered severe turbulence. The engineers solved the fog problem but unknowingly created another problem because of the tremendous amount of heat, energy, and turbulence created by the fires. Once re-engineered, the fog dispersal system provided an alternative to the too often practice of parachuting out of the plane just before letting it crash into the sea.[51] Maneuvering a large buoyant airship over a high-intensity fire would be fraught with the same problem. Increasing the height of the airship to avoid the turbulence would only result in the cascade of water evaporating before reaching the fire.

In 1930, the United States Forest Service experimented with the first containerized and free-fall water delivery systems. Beer kegs full of water were dropped onto a fire from a Ford Tri-motor airplane. The free-fall water drop involved delivering water through a long hose that hung freely from the plane.[52] The first operational application of water to suppress a fire occurred on September 9, 1950.[53] The Ontario Department of Lands and Forest dropped paper bags filled with water on a fire located north of Sault St. Marie. The results were favourable. What followed from these early primitive efforts was a remarkable period of ingenuity, perseverance, and technological development in the United States, Australia, and Canada.

A collaborative research program initiated in 1954 called Operation Firestop attested to the strong belief in the power of technology to conquer wildfire.[54] The United States Forest Service worked with the Los Angeles Fire Department, California Division of Forestry, United States Weather Bureau, Civil Defense Administration, the Marine Corps, and researchers at the University of California to evaluate news ways to suppress wildfires. Operation Firestop included the testing of water delivery systems that would eventually change how fires would be fought by air. Money was the only limiting factor.

Bigger budgets and bigger and faster air tankers replaced the need to build guards around settlements and other values-at-risk, and the need to burn out or set backfires. The increased firefighting horsepower also helped to ease the discomfort of the forest industry as they watched merchantable trees go up in flames from wildfires and burn out and backfire operations.

The aerial deployment of water and retardants by using rotary and fixed winged aircraft became the focus of fire suppression. Along the way of mixing and loading different retardants, testing new delivery systems, and providing specialized training, the art and science of fighting fire with fire were lost.

With the exception of the 1968 Slave Lake Fire in Alberta, the 1960s was a quiet decade with only seven fires each over 20,000 ha.[55] The following decade experienced an equally low level of fire activity with only eight fires over 20,000 ha.[56] Technology seemed to be winning the war against fire, and the public strongly identified this success with the use of aircraft.

The ability to effectively detect, initial attack, and suppress fires negated the need to prevent them. Education, enforcement, and engineering, the three pillars of prevention, took a back seat to aggressive suppression. In Alberta, the emphasis on suppression did not change until after the disastrous 1998 fire season. The following year, the prevention program was resurrected with the creation of the Wildfire Prevention Section, now the cornerstone of the wildfire management program in Alberta.[57]

We have gone full circle. Contemporary fire managers increasingly use fire to fight fire. They are also implementing fuel management strategies to mitigate the impact of wildfire. It is the same approach used in the early 1900s. In the mid-1930s, the residents of Anchorage, Alaska, concerned about nearby settler and railroad fires, decided to build a 183 m wide by 4,828 m long fireguard to protect their town.[58] Since Alaska did not establish their fire control organization until 1939, the community of Anchorage accepted responsibility and took action to defend itself from fires. They knew they lived and worked in a flammable forest environment. It was not a matter of if a fire would arrive but when it would arrive.

Learning how to work and live with fire is the goal of FireSmart programs in Canada and Firewise programs in the United States. These programs represent a shift in fire management strategy from focusing solely on aggressive suppression to including proactive and preventative management. The goal of FireSmart and Firewise programs is to plan and mitigate the wildfire hazards in areas where homes and communities are located within or near flammable vegetation.[59] When a home or building burns, the fire can spread to the adjacent forest. Likewise, a fire burning in the forest can spread rapidly and burn homes.

Reducing the risk of loss from these interface fires is a shared responsibility. The challenge is not in finding preventative solutions but in changing attitudes and behaviours to implement them. The reasons why people choose to live in what is referred to as the wildland–urban interface are the same reasons why they are at risk from fire.

Fighting fire is expensive and dangerous work. During very high winds, there is little any fire management organization can do to stop a large high-intensity fire. Six years after retirement, Eric Huestis talked to forest officers at an orientation course at the Alberta Forest Service Forest Technology School in Hinton, Alberta. While reflecting on the past, he succinctly summarized the problem:

> Then we progressed from the bulldozer to aircraft and we decided we were going to drop water, but we finally found out it was too damned expensive to drop water because a lot of it evaporated. We got about 30% of it on the ground after they had dropped it. The other 70% of it dried up. So we got this drilling mud from oil wells all over the country...We could pick it up pretty cheap...bentonite. Then we went to this gel which is better. We are using aircraft, and this is the world beater that is going to put out our fires. But, we are now learning that this isn't the world beater that is going to put out our fires. The aircraft can be used on a spot fire when it first starts. It can be used on a hot spot, and it can be used on one that's thrown over the main fire...but to put it on a big fire is just a waste of money. We are beginning to learn that.[60]

Hit fast. Hit hard. Keep 'em small. The fire management community achieves this mantra by the quick detection of a fire and the quick dispatch of initial attack resources. If the initial attack resources arrive to a fire when it is small, the probability of successfully containing it is high. Deciding on how many and what type of resources to send is often a challenge when the number of new fire starts overwhelms the available resources. Fire managers need to decide which fires get serviced first and which fires get put on hold. In some areas and under certain conditions, it may be desirable to allow a fire to occur.

Preparing before the fire arrives is the best strategy. Unfortunately, many homeowners still expect fire management agencies to protect their homes. This is a false expectation. On the morning of October 26, 2006, five United States Forest Service firefighters tried to protect a home that was surrounded by dry brush from an arson-set fire located 30 km northwest of Palm Springs, California.[61] Despite orders to relocate their engine once the Esperansa Fire entered the drainage, the firefighters stayed. The 30 m wall of flame pushed by the strong, dry Santa Ana winds was no match for the firefighters when they finally sought shelter. They all died. Three died on-site, one during transport to the hospital, and one later in the hospital. When a fire races across rugged, steep terrain, consuming tinder dry brush, nature is unequivocally in control. Only when the wind dies will firefighters succeed in gaining control. On June 5, 2009, 38-year-old Raymond Lee Oyler, the convicted arsonist who started the fire, was sentenced to death.[62]

The Esperanza wildfire burned 16,370 ha and destroyed 34 homes and 20 outbuildings.[63] It also trapped about 400 people by blocking the only road out of a recreational vehicle park located 153 km west of Los Angeles.[64] Access management is an important wildfire threat management consideration. When emergency vehicles enter and homeowners exit at the same time, and on the same road, problems inevitably occur.

In October 2003, southern California experienced its worst fire season in history. The easterly Santa Ana winds again blew hard, fanning 14 wildfires (Roblar 2, Pass, Grand Prix, Padua, Piru, Verdale, Happy, Old, Cedar, Simi, Paradise, Mountain, Otay, and Wellman).[65] On January 28, 2013, 31-year-old Rickie Lee Fowler was sentenced to death for starting the 36,940 ha Old Fire, which destroyed 993 homes and resulted in six fatalities. When the fire siege of 2003 ended, 3,710 homes were destroyed, and 24 people killed, including one firefighter.[66] Rick Andrea Tuttle, director of the California Department of Forestry and Fire Protection, and Jack Blackwell, regional forester for the Pacific Southwest Region of the United States Forest Service, co-authored an informative report called *California Fire Siege 2003—The Story*.[67] Tuttle and Blackwell indicated that similar fire events would happen again, hence their plea for more proactive planning. They were right. Fire and smoke filled the skies again in 2007, 2008, and 2009.

A home surrounded by flammable fuels burns with relative ease and speed in a wildfire. A home is source of fuel. If you own the home, you own the fuel, and if you own the fuel, you also own the responsibility for that fuel. Governments and insurance companies are now embracing this principle. It is no surprise that the 2003 Simi Fire in southern California had a lower loss rate of threatened homes than the Grand Prix, Old, Cedar, and Paradise Fires.[68] The Simi Fire was located in Ventura County where homeowners were required by law to establish and maintain a minimum 30.48 m (100 feet) defensible space around their homes.[69] Most homeowners who constructed their homes using less flammable materials, and managed the surrounding fuel, still had homes when they returned after the fire. Those homeowners who chose not to clear the brush around their homes or use less flammable construction materials often returned home after the fire only to find piles of rubble. In January 2005, the requirement to create a 30.48 m defensible space around homes and structures became a new state law in California. This legislated defensible space protects both homes and firefighters.[70]

In November 2008, 110 km/h Santa Ana winds fanned the Tea Fire burning in the hills above the towns of Montecito and Santa Barbara in southern California.[71] The oceanside town of Montecito is home to many celebrities. The median price of a home in Montecito is $2.9 million. Oprah Winfrey's 17 ha estate survived the fire but others did not. Wildfires do not much care who owns the home. The 785 ha Tea Fire in Santa Barabara County burned 210 residences.[72] In Corona and Orange County, the 4,239 ha Triangle Complex Fire burned 168 residences.[73] These fires destroyed multi-million dollar mansions, a mobile home park, and apartment buildings. The size and value of the residence are irrelevant. It's just a package of fuel.

Starting in July, wildfires made an early return throughout California in 2009. The month of August, which is normally outside the fire season, experienced several large wildfires. The largest fire, located north of Los Angeles, burned 64,983 ha. Two firefighters died when, in thick smoke, they drove their vehicle off a mountain road.[74]

Although firefighters encounter challenges when confronted with a firestorm pushed by 70 km/h dry winds, their overall fire suppression efforts are quite effective. In California, 97% of all wildfires are extinguished on the first day.[75] In Canada, 97% of all fires are contained at less than 200 ha in size.[76] The remaining 3% account for approximately 98% of the total area burned.[77] A small number of fires escape initial attack and become large fires. These firestorms consume forests, homes, and the suppression budgets of the fire management agencies.

The success of fire suppression over some 100 years has made it more difficult for firefighters today to continue to hold fire back. The removal of fire from pyrogenic landscapes yields undesirable and unnatural conditions down the road. Without fire, changes occur to the structure and composition of fire-dependent forests that have evolved with frequent low to moderate intensity surface fires. Over time, fuels build up, particularly, the ladder fuels. Suppression thus incubates fire.

The boreal forest in comparison is home to many high-intensity crown fires. Low- to moderate-intensity fires are relatively uncommon. The removal of small, medium, and large fires from the boreal landscape has changed the mosaic of forest age classes and patch sizes. The result at the landscape scale is a loss of natural fuel breaks created by fire, which, in turn, makes it more difficult for fire management agencies to prevent large fires from occurring later. Extinguishing fire in the boreal forest in essence removes firewalls or fire doors throughout the landscape. Many agencies now strategically create firewalls, particularly in protected areas, by using prescribed fire and, in some areas, mechanical treatment. If we do not do this, nature will manage the fuels for us.

The Canadian Council of Forest Ministers endorsed the development of a new vision for an innovative and integrated approach to managing fire in Canada by signing the Canadian Wildland Fire Strategy Declaration in 2005.[78] This ambitious initiative seeks to balance the good and the bad of fire. It recognizes the need to foster resilient communities and an empowered public, develop healthy and productive forest ecosystems, and incorporate modern and effective business practices. More needs to be done to manage public risk and expectations in the wildland–urban interface. The wildfires in California and Alberta are not going away. We, therefore, need to learn from fire and coexist with it. This is a paradigm shift. For fire management agencies, it is the big think.

CONCLUSION

The Big Think

> In Alberta, at least, forest fires [are] such a constant bed-fellow that there is no precedent to show what may happen when virgin stands are protected from fire for more than one rotation. It may not be too fantastic to consider the day when we shall wonder whether fire in some form at least, has not been a friend masquerading as a foe.
> —W.J. Bloomberg, "Fire and Spruce"[1]

While cruising for the Alberta Forest Service in the Eastern Rockies Forest Conservation Area during the late 1940s, William Bloomberg was in awe with the grandeur of the old spruce-fir stands but perplexed with the lack of spruce regeneration and the apparent suicidal tendency of these stands. Without fire, he believed the spruce component would eventually be eliminated. Bloomberg recognized fire as an agent of renewal; fire promoted the regeneration of spruce and thus ensured its perpetuation across the landscape.

Bloomberg's view of the ecological role of fire is now well understood. For this reason, fire management agencies acknowledge it is not ecologically desirable to remove all fire from the landscape. In Alberta, the Cordilleran Forests along the eastern slopes of the Rocky Mountains experience infrequent, large high-intensity fires, whereas the boreal forest is home to frequent, large high-intensity fires.

Over 90% of the area burned in Canada occurs in the boreal forest. The spatial distribution of the area burned is not uniform because the vegetation, topography, and weather vary across the boreal forest.[2] These differences have been classified into large areas with similar landforms, climate, and plant and animal communities, called ecozones. About 70% of the area burned in the boreal forest occurs in the Boreal Shield, Taiga Shield, and Taiga Plains Ecozones.[3] The fire–environment relationships and interactions are complex and non-stationary; explaining the subsequent spatial patterns of area burned is a challenge.[4]

Large fires (100,000–500,000 ha) are not uncommon in the boreal forest in Alaska and across Canada from Yukon to Quebec. From 1960 to 2011, 19 large fires occurred in Alberta. Fourteen of these fires were lightning-caused. Since 1919, five fires in Alberta exceeded 500,000 ha: Great Fire of 1919, 1982 Keane Fire, 1950 Chinchaga River Fire, 2002 House River Fire, and the 2011 Richardson Fire. The Northwest Territories has the biggest land and the biggest fires. These fires burn with little fanfare (1.1 million hectares Horn Plateau Fire in 1995, 857,600 ha Fire FS-029 in 1979, 624,883 ha Fire HY-049 in 1981, 553,680 ha Fire FS-078 in 1994, 477,000 ha Fire SM-002 in 1979, 438,000 ha Fire FS-024 in 1995, and the 428,100 ha Fire FS-012 in 1995).[5]

Twenty-five per cent of Wood Buffalo National Park, a 44,802 km² protected area located in northern Alberta and southern Northwest Territories, burned in 1981 (excluding water bodies and wetlands).[6] One of the most disastrous fire years in Canada occurred in 1989 when a record 3,567,947 ha burned in Manitoba.[7]

Historical fire sizes are often exaggerated because they include water bodies and unburned islands. Since fire management agencies only report areas burned within their jurisdiction, the reported size of a fire may also be misleading if it crossed provincial, territorial, or national park boundaries. The 2011 Richardson Fire, for example, burned approximately 707,000 ha. It also burned 46,793 ha in Wood Buffalo National Park and 1,683 ha in Saskatchewan.[8] When these areas and the water bodies and unburned islands are removed, the net burned area in Alberta is about 414,000 ha.

History shows the size of the Chinchaga River Fire is not itself unique. Its behaviour, too, is not out of the ordinary—short periods of rapid spread and long periods of slow spread. The smoke pall and its associated optical effects are, however, notable. Upper-level winds carried the thick layer of smoke entrapped between two inversion layers. This caused Black Sunday on September 24, 1950. A unique combination of smoke particle size and density resulted in the blue colouration of the sun and moon. The required volume of smoke did not come from the Chinchaga River Fire alone as first thought. Documented observations indicate the land of fire included four very large fires and many other fires of various sizes. This fire complex resulted in one of the largest burned areas—an estimated two million hectares—documented in North America.

The 1950 Chinchaga Firestorm in western Canada is a revealing story told many times before. Large fires occurred in the past, they occur today, and they will predictably occur in the future. The boreal forest is an ecosystem superbly adapted to fire. It is designed to coexist with fire. No other ecosystem holds and nurtures fire as a natural process like the boreal forest. In the deep organic layers, fire waits to fulfill its ecological role, not as an enemy but, as Bloomberg concluded, as a friend.

The indigenous peoples understood the ecological role of fire. They embraced it. Fire flowed surgically across the landscape. The indigenous peoples decided when and where to use fire based on whether the fire environment conditions allowed for its safe use. The early fire practitioners worked the entire forest to obtain different ecological services. Settlers, in comparison, forced fire on their quarter section of land—a piece of the fire environment they could not move. It was one location and for one purpose only. The settlers did not stop and wait for the weather conditions to change. They cleared, ploughed, and seeded the land as quickly as possible. Putting food on the table and clearing the land before the inspector arrived often meant burning debris piles and windrows in extreme and unsafe conditions. This resulted in escaped fires and problems for forest rangers like Frank LaFoy.

We need to understand fire as the indigenous peoples did. This means adapting and learning to live with fire. We too must embrace fire. Allowing all fire to roam freely across the boreal forest would be negligent. However, it is neither economically feasible nor ecologically desirable to eliminate fire from fire-dependent ecosystems.[9] Fire managers need to know which fires over time and space are keepers (wanted fire as an essential ecological process) and which fires are not (unwanted fire with negative impacts). When the fire load overwhelms the suppression response, informed decisions must be made. Distinguishing wanted fire from unwanted fire is also a useful input for land-use planning. This facilitates the alignment of decision making with land-use objectives and supports the use of prescribed fire.

Fire management agencies today cannot move homes and other values-at-risk to safer ground like the Keg River settlers did, but they can help develop resilient communities and empower residents to be informed, engaged, and better prepared to live and work in a fire-prone environment. Resilient communities strive to continuously implement initiatives (mitigation, preparedness, response, and recovery) to reduce the risk of loss from interface fires. This proactive approach, called FireSmart or Firewise, fosters effective and efficient fire management.

The Chinchaga River Fire and other 1950 fires in northeastern British Columbia and northwestern Alberta were human-caused fires. Lightning typically occurs later in the fire season when moisture and energy in the atmosphere support convection. Live and dead fuel moisture levels are at their lowest in the spring before green-up. This is the danger zone when even seemingly innocuous ignition agents such as smoldering organic matter packed on an all-terrain vehicle muffler, or a tree falling across a power line, can start a fast-moving fire when dry spring winds arrive.

On May 14, 2011, a cold, dry wind flowed into Alberta. A fire weather advisory issued the previous day warned of low relative humidity values and strong southeast winds. The spring stage was once again set. When the ignitions arrived the next day, a familiar scene quickly unfolded. Fierce winds blasted the lakeside Town of Slave Lake, located 200 km northwest of Edmonton, with embers, hot air, and smoke from a spring wildfire east of the town. The 10-minute average wind speed (recorded on the hour) on May 15 peaked at 57 km/h at 1700h at the Environment Canada weather station at the Slave Lake airport; wind gusts reached a maximum of 89 km/h.[10] Several nearby lookout towers reported wind gusts over 100 km/h. That night, despite the use of considerable suppression resources to stop its advance, Fire SWF-065-11 raced towards the town forcing the evacuation of all 7,000 people, who, using social media, watched horrific scenes of flames devouring their homes. This fire destroyed 428 single-family homes, 7 multi-family residences, and 19 non-residential buildings, including the town hall and downtown businesses.[11] Left homeless were 732 families. Fire SWF-056-11 southwest of Slave Lake also ran with the wind and challenged suppression efforts. These two fires destroyed an additional 56 single-family homes in the area surrounding the town.[12] A third fire (SWF-082-11) north of Slave Lake did not burn any structures. The three-fire complex known as the Flat Top Complex displaced almost 15,000 people from their residences and work places.

The 1950 Outlaw and Rat Fires caused an ember attack on the village of Keg River and throughout the Keg River prairie. The embers from these fires and also the Chinchaga River Fire likely started the Naylor Hills Fire, which destroyed Frank Jackson's sawmill. Like the 1950 fires, Fire SWF-065-11 caused a ferocious ember attack. The intense rain of embers caused the greatest havoc for the Town of Slave Lake. The buildings and firefighters had no chance. A third of the town vanished with ease and speed.

The Slave Lake Fires were not the only fires burning out of control in Alberta. The number of new fires accelerated from zero to 189 in five days. The resource demand to service this fire load exceeded the available suppression resources and their ability to contain the fires driven by gale force winds (89 km/h to 102 km/h winds are called whole gale or storm winds). These human-caused fires were preventable. Creating FireSmart communities and preventing human-caused fires will go a long way to help reduce losses from wildfires in the future. More importantly, living and working in the shadow of wildfire is a shared responsibility.

Supersized fires are called mega fires, a term coined by Jerry Williams, former chief forester with the United States Forest Service.[13] Mega fires are extreme fires—bigger, hotter, and faster, and hence more destructive, dangerous, and expensive. They exceed control efforts until the weather or fuel changes. The estimated $742 million in damages from the Slave Lake Fires activated the second-highest natural disaster insurance claim in Canada.[14] The Alberta government provided $289 million to assist the affected communities with their rebuilding efforts.[15] Firefighting expenditures pushed the total cost over one billion dollars— all because of one short, extreme weather event.

Mega fires are a concern because of climate change. The Intergovernmental Panel on Climate Change Fourth Assessment Report reported a 0.76 °C to 0.95 °C increase in temperature from the 1850–1899 period to the 2001–2005 period.[16] Global climate data sets (National Oceanic and Atmospheric Administration National Climatic Data Center, National Aeronautics and Space Administration Goddard Institute for Space Studies, and the Hadley Centre for the UK Met Office) show a global warming trend of about 0.6 °C to 0.8 °C over the past 100 years.[17] Nine of the ten warmest years have occurred since 2000. The earth is warming and its weather is getting more extreme. This volatility includes drought, floods, tornadoes, hurricanes, and fire.

Climate and climate change processes, including the interactions between land cover, clouds, and oceans, and their influence on the world's energy balance, are complex. Oceans absorb about one-third of the carbon humans release into the atmosphere, but their ability to absorb more carbon may be lessening.[18] Forests and fires are also wildcards. Forests absorb one-third of the carbon dioxide emitted by humans.[19] Fire can quickly flip forests from sink to source. Carbon dioxide is released directly during combustion and indirectly through the decomposition of the fire-killed vegetation. The young, vigorous successional forest stands will in 20 or 30 years sequester more carbon.

Fire decides whether our boreal forests become carbon sinks or sources. This essentially makes fire managers carbon managers.

Our understanding of the processes that contribute to climate change is itself changing as new research is published. For example, scientists studying the role of black carbon (soot) suggest these particles are the second-largest contributor to global warming—twice the level of previous estimates. The surprising results of a comprehensive four-year, black carbon study published in the *Journal of Geophysical Research: Atmospheres* could shift the focus from managing carbon dioxide emissions to managing the black carbon emissions.[20] Forest fire management can potentially be an important black carbon mitigation strategy.

We can expect more extreme wind events and longer fire seasons (starting earlier and ending longer). No longer can we assume the usual June rains or September season-ending rain events will occur. Fire loads will increase, and, perhaps the most alarming forecast, the synchronicity of bad fire seasons (east and west in Canada, North America, and Europe) will cause national and global resource crunches.

Climate change and a weather-altering pattern called El Niño have opened the door for fire to go places it has not been to before. El Niño is a cyclic warm Pacific Ocean phase. The associated high surface pressure and warm ocean temperature cause very dry conditions. During El Niño years, rain forests influenced by the Pacific Ocean become unusually aglow with fire. A strong El Niño occurred in 1997–98. Fires in Indonesia burned an estimated 750,000 ha, according to government officials. The Indonesia Forum for Environment suggested over 1,700,000 ha burned.[21]

Compared to the large fires in the boreal forests of Canada, Russia, and Alaska, the land-clearing broadcast fires in the rain forests are small but many. In 2007, NASA's Terra Satellite detected 50,729 fires in the Brazilian Amazon from July to September. NASA's Aqua Satellite detected 72,329 fires.[22] This deforestation is ecologically destructive because the rain forest is adapted to rain, not fire.

Whether from the 1950 Chinchaga Firestorm or the 2011 Slave Lake Fires, learning from the past can help fire management agencies manage uncertainty in a changing climate. It speaks of what the future may look like. In 1950, fire took advantage of a regional drought in northwestern Alberta. Allowed to burn freely, the Chinchaga River

Fire burned throughout the fire season until rain and snow arrived in early October. Big fires are a part of the boreal forest. Another Chinchaga River Fire occurred in Alberta in 2011. This was the Richardson Fire. Again, fire took advantage of a regional drought, this time in northeastern Alberta. With suppression resources committed to high-priority fires, the human-caused Richardson Fire was allowed to burn. When the expected June rains failed to arrive, it continued to spread. Suppression efforts focused successfully on protecting the values-at-risk and waiting for periods of slow fire spread to build guards.

Statistics Canada reported a decrease in oil and gas extraction of 4.2% in May 2011. The shutdown of pipelines and plants and the evacuation of large work camps because of the Richardson Fire contributed to the drop in energy production. This decrease was significant enough to also decrease Canada's real gross domestic product in May by 0.3%.[23]

Big fires generate big questions. Big questions then generate big responses. The big response from the 1950 Chinchaga Firestorm was the removal of the 16 km firefighting limit by the Forest Service on September 16, 1952. Policy will continue to be an important instrument to adapt, but science and technology will increasingly help fire managers unravel and reconsider what is possible in a complex, changing world. Contemporary fire managers certainly have more tools than Frank LaFoy did to provide decision support, but bigger and faster fires require bigger and faster tools. This includes improved fire weather, fire behaviour, fire occurrence, and resource-demand forecasting. More wildfires and prescribed fires across the landscape require fire management agencies to sustain and, where possible, strengthen their fire suppression capability.

In a 1925 Christmas Eve memorandum to his director, Ernest Finlayson, Jim Wright, a civil engineer with the Department of the Interior Forestry Branch, planted the seed for the initiation of an operational fire research program that led to the development of the Canadian Forest Fire Danger Rating System.[24] Wright understood the fundamental relationships between weather, fuel moisture, and fire behaviour. With support from the Dominion Forest Service, Wright began a 25-year research journey to apply his engineering skills to quantify these relationships. Wright developed a system of fire hazard measurement based on his research on the effect of fuel moisture on fire flammability.[25] Herbert Beall joined the research team in 1928 working as a summer student for Wright. The Forest Service quickly recognized Beall's potential and hired him full time in 1932. Together, Wright and Beall crafted the first generation fire danger rating system in Canada.

In 1930, Alberta assumed responsibility for forest protection under the Transfer of Resources Act.[26] Wright's fire research program nevertheless continued at the Petawawa Forest Experiment Station at Petawawa, Ontario. Other experimental stations opened in Quebec, New Brunswick, Manitoba, Saskatchewan, and Alberta during the 1930s. The collective research from this network provided the scientific basis for understanding how fires start and spread.

Wright installed a state-of-the-art weather station at Petawawa to record weather observations twice a day. It included a maximum and minimum thermometer, wet and dry bulb thermometer, and an Assman hygrometer to measure relative humidity. A thermohygrograph recorded temperature and relative humidity. Pressure was measured and recorded using a barometer and barograph. An automatic electric anemobiograph measured and recorded wind speed and wind direction. The amount of precipitation was measured using an automatic recording rain gauge. Several instruments were used to measure the amount of sunshine and evaporation, including a pyrometer designed by Wright.

Wright published fire hazard tables in 1931 specific for white pine sites.[27] These tables used daily weather readings to produce flammability indices for five hazard levels. Wright acknowledged these tables could not be applied to other regions with different climate and fuel types. By 1948, the tables, then referred to as fire danger tables, expanded to include a range of forest and fuel types.[28]

The Meteorological Division in the Department of Transport provided fire hazard weather forecasting for forest protection agencies across Canada. These forecasts, sent by telegraph, allowed for the calculation of forecasted Wright Fire Danger Indices. In 1950, estimates of forest fire danger based primarily on weather observations were made each day for high-priority forested areas in Canada. Frank LaFoy's district was not a high-priority area. With fewer fires, but more people, the Eastern Rockies Forest Conservation Area (ERFCA) topped the priority list.

During the 1950 fire season, the ERFCA tested and compared the Wright System with fuel moisture sticks installed at their weather station.[29] These pine dowels emulated forest fuels. Their initial dry weight subtracted from the weight in ambient conditions in the forest yielded the per cent moisture content of the forest fuels. The Northern Alberta Forestry District (NAFD) decided instead to test a modified form of the Gisborne Fire Danger Meter and compare it with actual fire occurrences and subjective assessments of fire danger made by forest rangers.[30] Collaborative research between the NAFD and ERFCA got off to slow start.

Despite not having any fuel moisture sticks or meters at Keg River, LaFoy knew when the snow and ice left the Chinchaga River Valley. He also knew how many days passed since it last rained and the relative moisture level of the forest fuels. LaFoy knew this because he spent most of his time in the forest. Experience is critical when predicting how fire will behave. Australian fire researcher Neil Burrows concluded, "Expectations of how a fire should behave are based largely on experience, and to a lesser extent, on fire behaviour guides."[31] Although experience is important, fire managers today cannot escape the increasing application of powerful decision support tools to confront an increasingly complex fire environment.

Wright and Beall saw value in weather forecasting: "A true picture of the fire situation at a given moment is of great value, but foreknowledge of how the picture is likely to appear in a few hours or days can be still more valuable, since it permits of intelligent planning to meet the changing situation. It is here that the meteorologist can render valuable aid with his science of weather forecasting."[32]

A new vision for wildfire management called the Canadian Wildland Fire Strategy recognizes the need to improve fire weather and fire behaviour forecasting.[33] These forecasting tools would have helped LaFoy better prepare for the arrival of the big wind. Using a two-way radio, Ray Ross, manager at the Keg River Hudson's Bay trading post talked to his colleagues at the Hudson's Bay trading post at Fort Nelson, who reported very strong winds. The big wind was moving east but LaFoy did not know when it would hit Keg River.[34] Had he known, he could have provided the settlers with more notification time to prepare and, if necessary, evacuate.

The lack of suppression resources and the 16 km firefighting limit constrained what LaFoy could do. He received strict orders not to dispatch a firefighting crew to the Chinchaga River Fire. When the Naylor Hills Fire started within 16 km of the McKenzie Highway, fire suppression efforts still did not commence until LaFoy obtained approval from the Division Office in Peace River.[35]

Not all of the homesteads in the Keg River prairie survived unscathed from the 1950 firestorm. Some buildings, equipment, and crops were destroyed and damaged, but no one died. When LaFoy saved the village by back burning, he was too tired to chant the mantra, "The Alberta Forest Service gets things done!" The singing would be heard many years later.

The lessons learned from the 1950 Chinchaga Firestorm in western Canada and other large fires in the past help guide fire managers into the future. They speak loudly of the need to embrace fire and shift from response to prevention and preparedness. A strong suppression capability, though essential, cannot solve the problem on its own. It is not possible to stop all fires at all times. Since agencies become challenged during worst-case fire events, the sharing of resources across Canada will become increasingly important.

Lastly, the lessons learned speak of great men like Frank LaFoy, Eric Huestis, and Frank Jackson. These three individuals were leaders and, in their own way, visionaries. Where there are great fires, there are great men. The new vision for fire management in Canada now calls for today's leaders to listen to the 1950 Chinchaga Firestorm and lead wildfire management into the future where people, fire, and forests coexist, and fire science and technology help light the path forward.

NOTES

FOREWORD

1   Fay H. Johnston et al., "Estimated Global Mortality Attributable to Smoke from Landscape Fires," *Environmental Health Perspectives* 120, no. 5 (2012): 695–701.

INTRODUCTION

1   Natural Resources Canada, *The State of Canada's Forests Annual Report 2010* (Ottawa: Natural Resources Canada, Canadian Forest Service, 2010).
2   Data from 1990 to 2009 were obtained from the National Forestry Database website (accessed March 27, 2011): http://nfdp.ccfm.org/data/compendium/html/comp_32e.html. Data for 2010 were obtained from the Canadian Interagency Forest Fire Centre (CIFFC) website (accessed March 27, 2011): http:www.ciffc.ca/firewire/current.php.
3   Natural Resources Canada, *The Atlas of Canada, Boreal Forest*, accessed March 26, 2011, http://atlas.nrcan.gc.ca/site/english/learningresources/theme_modules/boreal-forest/index.html.
4   Table 3.2. Number of Forest Fires in the Intensive and Limited Protection Zones by Response Category, Cause Class, and Province/Territory/Agency, 1990–2010,

Canadian Council of Forest Ministers, National Forestry Database Program, accessed March 26, 2011, http://nfdp.ccfm.org/data/compendium/html/comp_32e.html.

5   B.J. Stocks, J.G. Goldammer, and L. Kondrashov, "Forest Fires and Fire Management in the Circumboreal Zone: Past Trends and Future Uncertainties". (IMFN discussion paper no. 01, International Model Forest Network Secretariat, Natural Resources Canada, Ottawa, 2008).

6   M.J. Apps et al., "Boreal Forests and Tundra," *Water, Air, and Soil Pollution* 70 (1993): 39–53.

7   T.F. Blefgen, *Report of the Director of Forestry. Annual Report of the Department of Lands and Mines of the Province of Alberta for the Fiscal Year Ended March 31st 1943* (Edmonton: King's Printer, 1944), 36.

8   S. Welke and J. Fyles, "Organic Matter: Does It Matter?" (SFMN research note series no. 3, Sustainable Forest Management Network, University of Alberta, Edmonton, 2005).

9   R.T. Watson et al., eds. *Land Use, Land-Use Change, and Forestry. Special Report of the Intergovernmental Panel on Climate Change* (IPCC) (Cambridge: Cambridge University Press, 2000).

10   R.W. Mutch, "Wildland Fires and Ecosystems—A Hypothesis," *Ecology* 51 (1970): 1046–51.

11   H.J. Gratkowski, "Heat As a Factor in Germination of Seeds of Ceanothus velutinus var. laevigatus T. & G." (PhD diss., Oregon State University, Corvallis, 1962).

12   C.G. Wells et al., "Effects of Fire on Soil, A State-of-Knowledge Review" (general technical report WO-7, USDA Forest Service, Washington, DC, 1979); L.F. DeBano, "Chaparral Soils," in *Symposium on Living with the Chaparral: Proceedings*, ed. Murray Rosenthal (San Francisco: Sierra Club, 1974), 19–26; F.R. Steward, "Heat Penetration in Soils beneath a Spreading Fire" (unpublished paper, Intermountain Forest and Range Experiment Station, USDA Forest Service, Missoula, MT, 1989).

13   W.G. Evans, "Perception of Infrared Radiation from Forest Fires by Melanophila Acuminata de Geer (Buprestidae, Coleoptera)," *Ecology* 47 (1966): 1061–65.

14   S. Hart, "Beetlemania: An Attraction to Fire," *BioScience* 48, no. 1 (1998): 3–5.

15   W.T. McDonough, "Quaking Aspen-Seed Germination and Early Seedling Growth" (research paper, INT-234, Intermountain Forest and Range Experiment Station, Ogden, UT, 1979).

16   J.S. Maini, "Silvics and Ecology of Populus in Canada," in *Growth and Utilization of Poplars in Canada*, eds. J.S. Maini and J.H. Cayford (Ottawa: Canada Department of Forestry and Rural Development, 1968), 20–69.

17   H.T. Lewis, "A Time for Burning" (occasional publication no. 17, Boreal Institute for Northern Studies, University of Alberta, Edmonton, 1982).

18   Ibid.

19   Ibid., 1.

20   Ibid., 19.

21   Orest T. Martynowych, "The Ukrainian Bloc Settlement in East Central Alberta, 1890–1930: A History" (occasional paper no. 10, Historic Sites Service, Alberta Culture, Edmonton, 1985), 118–19.

22   Frank LaFoy, "The 1950 Chinchaga River Fire. Excerpts from an Interview with Frank L. LaFoy, Former Forest Ranger," by John Frank, undergraduate forestry student, Winter 1977 (directed study for Peter J. Murphy), eds. Peter J. Murphy and

Cordy Tymstra and transcribed by Judy Jacobs (Faculty of Agriculture and Forestry, University of Alberta, Edmonton, Fall 1985), 2.

23   Tom Philip, "Pioneering Peace Medicine: How Dr. Mary Percy Jackson Conquered the Wild and Healed the Sick," *Alberta Report*, December 26, 1983, 30–33.

24   Frank Jackson, *A Candle in the Grub Box: A Struggle for Survival in the Northern Wilderness. The Story of Frank Jackson As Told to and Transcribed by Sheila Douglass* (Victoria, BC: Shires Books, 1977), 83–84; a one-page recollection of events written by Frank Jackson and entitled "Frank Jackson's story, written in 1975. Page 756." This one-page summary was provided to Peter J. Murphy by Mary Percy Jackson with her letter to him in Mary Percy Jackson to Peter J. Murphy (associate dean, Forestry, University of Alberta, Edmonton), February 29, 1984, letter in author's possession.

25   Mary Percy Jackson to Peter J. Murphy (associate dean, Forestry, University of Alberta, Edmonton), February 29, 1984, letter in author's possession.

26   *The Gateway* (University of Alberta), "Varsity Win from Sask. in Varsity Soccer Final," November 14, 1922, 5, http://peel.library.ualberta.ca/newspapers/GAT/.

27   cbc-Radio, *Round Up*, featuring Gerrard Faye, originally broadcast September 28, 1950, cbc Digital Archives, accessed March 28, 2011, http://www.cbc.ca/archives/categories/environment/natural-disasters/blue-sun-shines-over-england.html.

28   Letter and three-page enclosure from Peter Murphy (associate dean, Forestry, Faculty of Agriculture and Forestry, University of Alberta, Edmonton) to Peter Fuglem (Pacific Forest Research Centre, Canadian Forestry Service, Victoria, BC), October 28, 1982. The enclosure included a copy of a photograph of Ron Hammerstedt's composite map and two tables summarizing the areas burned of the fires related and unrelated to the Chinchaga River Fire during the September 20–21 period.

29   P.J. Murphy, "Fire Research at the University of Alberta," in *Proceedings of Fire Ecology in Resource Management Workshop, Dec. 6–7, 1977, Edmonton*, comp. D.E. Dubé (information report NOR-X-210, Canadian Forest Service, Northern Forest Research Centre, Edmonton, 1978), 90–91.

30   Cordy Tymstra, "Fire Behaviour Study—Chinchaga Fire," FOR 451: Intermediate Forestry Problems course, University of Alberta, Edmonton, 1985.

31   P.J. Murphy, interview with the author, Edmonton, November 24, 2010.

32   Associated Press, "Smoke, Pall, Chill Mark Weather Map," *Reno Evening Gazette*, September 25, 1950, Nevada State Library and Archives.

33   Partners in Protection, *FireSmart: Protecting Your Community from Wildfire*, 2nd ed. (Edmonton: Partners in Protection, 2003). FireSmart is a proactive approach to living and working in the shadow of wildfire. It means being prepared when wildfires arrive because more firefighting aircraft combined with aggressive initial attack alone cannot stop all wildfires all the time.

ONE   THE LOST BC FIRE

1   Keg River History Book Committee, *"Way Out Here": History of Carcajou, Chinchaga, Keg River, Paddle Prairie, Twin Lakes* (Altona, MB: Friesen Printers, 1994).

2   Ibid.

3   Ibid.

4   Ibid.

5   T.F. Blefgen, *Report of the Director of Forestry. Annual Report of the Department of Lands and Mines of the Province of Alberta for the Fiscal Year Ended March 31st 1935* (Edmonton: King's Printer, 1936), 73–102.

6   T.F. Blefgen, *Report of the Director of Forestry. Annual Report of the Department of Lands and Mines of the Province of Alberta for the Fiscal Year Ended March 31st 1937* (Edmonton: King's Printer, 1938), 47–76.

7   Ibid.

8   Ibid.

9   W.R. Hanson, *History of the Eastern Rockies Forest Conservation Board, 1947–1973* (Calgary: Eastern Rockies Forest Conservation Board, 1973).

10   Peter J. Murphy, "'Following the Base of the Foothills': Tracing the Boundaries of Jasper Park and Its Adjacent Rocky Mountain Forest Reserve," in *Culturing Wilderness in Jasper National Park: Studies in Two Centuries of Human History in the Upper Athabasca River Watershed*, ed. I.S. MacLaren (Edmonton: University of Alberta Press, 2007), 71–121.

11   Ibid.

12   Memorandum from F. Neilson (chief timber inspector, Alberta Forest Service) to all timber inspectors, June 4, 1945, Provincial Archives of Alberta.

13   E.S. Huestis, *Report of the Director of Forestry. Annual Report of the Department of Lands and Mines of the Province of Alberta for the Fiscal Year Ended March 31st 1949* (Edmonton: King's Printer, 1950), 18–55.

14   Ibid.

15   E.S. Huestis, *Report of the Director of Forestry. Annual Report of the Department of Lands and Mines of the Province of Alberta for the Fiscal Year Ended March 31st 1950* (Edmonton: King's Printer, 1951), 26–72.

16   E.S. Huestis, *Report of the Assistant Director of Forestry. Annual Report of the Department of Lands and Mines of the Province of Alberta for the Fiscal Year Ended March 31st 1948* (Edmonton: King's Printer, 1949), 31–73.

17   Huestis, *Report of the Director of Forestry ... for the Fiscal Year Ended March 31st 1949*, 18–55.

18   A cache is a small building used to store firefighting equipment for use when needed.

19   Huestis, *Report of the Director of Forestry ... for the Fiscal Year Ended March 31st 1950*, 26–72.

20   LaFoy, "The 1950 Chinchaga River Fire," 2.

21   J. McKee, "Fire Protection in British Columbia," in *Proceedings of the 55th Session of the Pacific Logging Congress. Loggers Handbook*, vol. 25 (Portland, OR: Pacific Logging Congress, 1965), 94–96.

22   British Columbia Department of Lands and Forests, *Whisp Fire Report* (Fort St. John: Forest Service, 1950).

23   Memorandum from E.S. Huestis (acting assistant director of Forestry) to Mr. J. Harvie (deputy minister, Department of Lands and Mines), signed by E.S. Huestis and signed and approved by J. Harvie, March 29, 1945, Government of the Province of Alberta, file no. Head Office, Fires (General).

24   Alberta Department of Lands and Mines, *Boundry Fire Report* (Peace River: Forest Service, 1950).

25   Ibid.

26   Base map with the Chinchaga River Fire perimeter and LaFoy's handwritten notes, in author's possession.

27   John Parminter, *1950 Fire Report Summaries* (Victoria: British Columbia Forest Service, 1981).

28   Eric LaFoy (Frank LaFoy's son), interview with Cordy Tymstra and Peter Murphy, Boyle, AB, June 21, 2011.

29   Government of Alberta, *The Alberta Story: An Authentic Report on Alberta's Progress, 1935–1952* (Edmonton: Publicity Bureau, Department of Economic Affairs, 1952).

30   Ibid.

31   C.F. Platt, "Fire Suppression," in *Complete Set of Papers for the Alberta First Natural Resources Conference, Theme: Inventory* (Edmonton: Government of Alberta, 1959).

32   Ibid.

33   Alberta Department of Lands and Mines, *No. 11, Orders in Council, 1948* (Edmonton: OIC 825/48, Department of Lands and Mines, 1948).

34   Ibid.

35   LaFoy, "The 1950 Chinchaga River Fire," 6.

36   Ibid.

37   Ibid.

38   Ibid.

39   Memorandum from E.S. Fellows (chief forester) to the chairman and members of the Eastern Rockies Forest Conservation Board, October 5, 1951. This memorandum includes an eight-page report and recommendations on a fire-detection system for the Forest Reserve.

40   Ibid.

41   Ibid.

42   LaFoy, "The 1950 Chinchaga River Fire."

43   T.F. Blefgen, *Report of the Director of Forestry. Annual Report of the Department of Lands and Mines of the Province of Alberta for the Fiscal Year Ended March 31st 1955* (Edmonton: King's Printer, 1956), 26–72.

44   Memorandum from E.S. Fellows (chief forester) to the chairman and members of the Eastern Rockies Forest Conservation Board, October 5, 1951.

45   Ibid.

46   Platt, "Fire Suppression."

47   Ibid.

48   Ibid.

49   Eastern Rockies Forest Conservation Board, *Annual Report of the Eastern Rockies Forest Conservation Board for the Fiscal Year 1949–50* (Calgary: Eastern Rockies Forest Conservation Board, 1950), 11.

50   Ibid., 12.

51   Eastern Rockies Forest Conservation Board, *Annual Report of the Eastern Rockies Forest Conservation Board for the Fiscal Year 1950–51* (Calgary: Eastern Rockies Forest Conservation Board, 1951).

52   Eastern Rockies Forest Conservation Board, *Annual Report...for the Fiscal Year 1949–50*.

53   LaFoy, "The 1950 Chinchaga River Fire."

54   Ibid.

55   Shirlee Smith Matheson, *Lost: True Stories of Canadian Aviation Tragedies* (Calgary, AB: Fifth House Ltd., 2005).

56   LaFoy, "The 1950 Chinchaga River Fire," 29.
57   Letter to Mr. Kirk Underschultz (Edmonton) from Frank Jackson (Keg River, AB), August 2, 1977, letter in author's possession.
58   Ibid.
59   LaFoy, "The 1950 Chinchaga River Fire,"#10.
60   Letter to Mr. Kirk Underschultz (Edmonton) from A.H. Edgecombe (senior fire control instructor, Forest Technology School, Hinton, AB), August 26, 1977, letter in author's possession; Hinton Training Centre archival records, Hinton, AB, Alberta Energy and Natural Resources, file 40-1, copy of record provided to the author by P.J. Murphy.
61   Tim Klein, personal communication (email) with the author, October 31, 2005.
62   LaFoy, "The 1950 Chinchaga River Fire," 25.
63   *Battle River Herald*, "Trappers Escape Northern Bush Fires," September 28, 1950.
64   Canadian Pulp and Paper Association, *Forest Fire Fighters Guide* (Montreal: Canadian Pulp and Paper Association, Woodlands Section, 1949), 31–32.
65   Ibid., 44–45.
66   Donna Enger, personal communication with the author, March 2007.
67   LaFoy, "The 1950 Chinchaga River Fire," 18.
68   Ibid., 31.
69   Ibid., 18, 19.
70   The exact location of northwest quarter section was Township 101, Range 24, Section 33, W5.
71   *Battle River Herald*, "Keg River News," October 12, 1950.
72   Letter from Mary Percy Jackson to Peter J. Murphy (associate dean, Forestry, University of Alberta, Edmonton), February 29, 1984, letter in author's possession.
73   Jackson, *A Candle in the Grub Box*.
74   Ibid.
75   Letter from Mary Percy Jackson to Peter J. Murphy (associate dean, Forestry, University of Alberta, Edmonton), February 29, 1984, letter in author's possession.
76   Alberta Department of Lands and Forests, *Rat Fire Report* (Peace River: Forest Service, 1950).
77   Ibid., 2.
78   *Edmonton Journal*, "Swept by Fire: Farm Homes Destroyed in Wanham District," September 21, 1950.
79   Ibid.
80   *Edmonton Journal*, "700 Battle Bush Fires," September 23, 1950.
81   Ibid.
82   Birch Hills Historical Society, *Grooming the Grizzly: A History of Wanham and Area*, ed. W. Tansen (Winnipeg, MB: Inter-Collegiate Press, 1982).
83   *Edmonton Journal*, "Swept by Fire: Farm Homes Destroyed in Wanham District," September 21, 1950.
84   Birch Hills Historical Society, *Grooming the Grizzly*.
85   Alberta Department of Lands and Forests, *Outlaw Fire Report* (Peace River: Forest Service, 1950).
86   Alberta Department of Lands and Forests, *Naylor Hills Fire Report* (Peace River: Forest Service, 1950), 2.
87   Ibid.

88   A one-page recollection of events written by Frank Jackson and entitled "Frank Jackson's story, written in 1975. Page 756." This one-page summary was provided to Peter J. Murphy by Mary Percy Jackson with her letter to him.

89   Alberta Department of Lands and Forests, *Naylor Hills Fire Report*, 2.

90   Alberta Department of Lands and Forests, *Willow Swamp Fire Report* (Peace River: Forest Service, 1950).

91   *Edmonton Journal*, "700 Battle Bush Fires," September 23, 1950.

92   Ibid.

93   *Edmonton Bulletin*, "100 Bush Fires Ravaging North," September 25, 1950.

94   Ibid.

95   Ibid.

96   *Globe and Mail*, "100 Blazes Unchecked in B.C., Alberta Timber," September 26, 1950.

97   Ibid.

98   *Edmonton Bulletin*, "100 Bush Fires Ravaging North," September 25, 1950.

99   Parliament of Victoria, *2009 Victorian Bushfires Royal Commission: Final Report Summary* (Victoria, AU: Government Printer for the State of Victoria, 2009).

100   J. Cowan, "Wiped Out: Town Destroyed by Killer Fires," ABC *News*, February 8, 2009, http://www.abc.au/news/stories/2009/02/08/2485378.htm.

101   Letter from Mary Percy Jackson to Peter J. Murphy (associate dean, Forestry, University of Alberta, Edmonton), February 29, 1984, letter in author's possession.

102   Mary Percy Jackson and Cornelia Lehn, *The Homemade Brass Plate: The Story of a Pioneer Doctor in Northern Alberta as Told to Cornelia Lehn* (Sardis, BC: Cedar-Cott Enterprise, 1988).

103   Tom Philip, "Pioneering Peace Medicine: How Dr. Mary Percy Jackson Conquered the Wild and Healed the Sick," *Alberta Report*, December 26, 1983, 30–33.

104   Jackson and Lehn, *Homemade Brass Plate*, 110.

105   Mary Percy Jackson, *Suitable for the Wilds: Letters from Northern Alberta 1929–1931* (Calgary, AB: University of Calgary Press, 2006).

106   Keg River History Book Committee, *"Way Out Here."*

107   David W. Leonard, *The Last Great West: The Agricultural Settlement of the Peace River County to 1914* (Calgary, AB: Detselig Enterprises Ltd., 2005).

108   Ibid.

109   Keg River History Book Committee, *"Way Out Here."*

110   Ibid.

111   Jackson, *A Candle in the Grub Box.*

112   Keg River History Book Committee, *"Way Out Here."*

113   Jackson and Lehn, *Homemade Brass Plate.*

114   Tom Philip, "Pioneering Peace Medicine," *Alberta Report*, December 26, 1983, 30–33.

115   Anne Vos, "History of Dr. Mary Percy Jackson," *Dr. Mary Percy Jackson*, accessed June 11, 2011, http://www.drmaryjackson.com/maryjackson.html.

116   Jackson, *A Candle in the Grub Box.*

117   Anne Vos, "History of Dr. Mary Percy Jackson," *Dr. Mary Percy Jackson*, accessed June 11, 2011, http://www.drmaryjackson.com/maryjackson.html.

118   Letter from Mary Percy Jackson to Peter J. Murphy (associate dean, Forestry, University of Alberta, Edmonton), February 29, 1984, letter in author's possession.

119   Ibid.

120   Keg River History Book Committee, *"Way Out Here."*

121   LaFoy, "The 1950 Chinchaga River Fire."

122 E.E. Ballantyne, "Rabies Control in Alberta," *Canadian Journal of Comparative Medicine and Veterinary Science* 20, no. 1 (1956): 21–30.

123 Jack Grant (forest ranger, chief ranger, aircraft dispatcher), interview with P.J. Murphy, Edmonton, January 26, 1999.

124 Canada, Department of the Interior, "Annual Report of the Department of the Interior for the Year 1897" (sessional papers, no. 13. A., Department of the Interior, Ottawa, 1898).

TWO BLACK SUNDAY

1 Betty Rhodes, post to "Erie Legends," *goerie.com*, June 17, 2000, http://www.angelfire.com/pe/rcmatteson/sunday.html. Betty Rhodes's posts and the subsequent thread cited in this chapter are no longer available on the *GoErie* website but have been archived at http://the-red-thread.net/sunday.html.

2 Ibid.

3 "Interesting Highlights of the Warren Area," accessed October 16, 2014, http://lynnmariebriggs.com/warren.html.

4 N.P. Carlson, "One Sunday Afternoon in 1950 the Sun Went Out," *Post-Journal* (Jamestown, NY), January 3, 1987.

5 *New York Times*, "Forest Fires Cast Pall on Northeast: Canadian Drift 600 Miles Long Darkens Wide Areas and Arouses 'Atom' Fears," September 25, 1950.

6 William L. Miller, reply to Betty Rhodes's post to "Erie Legends," *goerie.com*, accessed February 18, 2002, http://www.angelfire.com/pe/rcmatteson/sunday.html.

7 K.C. Harper, *Weather and Climate: Decade by Decade* (New York: Facts on File, 2007).

8 Jerry E. Smith, *Weather Warfare: The Military's Plan to Draft Mother Nature* (Kempton, IL: Adventures Unlimited Press, 2006).

9 Larry Knickerbocker, reply to Betty Rhodes's post to "Erie Legends," *goerie. com*, accessed February 18, 2002, http://www.angelfire.com/pe/rcmatteson/sunday.html.

10 *New York Times*, "Forest Fires Cast Pall on Northeast. Canadian Drift 600 Miles Long Darkens Wide Areas and Arouses 'Atom' Fears," September 25, 1950.

11 E.M. Elsley, "Alberta Forest-Fire Smoke—24 September 1950," *Weather* 6, no. 1 (1951): 22–26.

12 *Reno Evening* Gazette, "16 Men Survive Crash of Bomber: Rescue Is Begun In Wilderness," September 25, 1950.

13 *Globe and Mail*, "Alberta Smoke Covers Toronto," September 25, 1950.

14 Cliff Shilling, reply to Betty Rhodes's post to "Erie Legends," *goerie.com*, accessed February 18, 2002, http://www.angelfire.com/pe/rcmatteson/sunday.html.

15 *New York Times*, "Forest Fires Cast Pall on Northeast: Canadian Drift 600 Miles Long Darkens Wide Areas and Arouses 'Atom' Fears," September 25, 1950.

16 Ibid.

17 *Toronto Star*, "Postponed Due to Smog? Woman Asks, but Eclipse on Dot, and Nobody Shot," September 26, 1950.

18 N.P. Carlson, "One Sunday Afternoon in 1950 the Sun Went Out," *Post-Journal* (Jamestown, NY), January 3, 1987.

19   Ibid.
20   Jayne B. Mortenson, personal communication (email) with author, July 6, 2001.
21   Elsley, "Alberta Forest-Fire Smoke," 22–26.
22   C.D. Smith, "The Widespread Smoke Layer from Canadian Forest Fires during Late September 1950," *Monthly Weather Review*, September 1950, 180–84.
23   G.A. Bull, "Blue Sun and Moon," *Marine Observer* 21, no. 153 (1951): 167–69.
24   Peter Murphy and Cordy Tymstra, "The 1950 Chinchaga River Fire in the Peace River Region of British Columbia/Alberta: Preliminary Results of Simulating Forward Spread Distances," in *Proceedings of the Third Western Region Fire Weather Committee Scientific and Technical Seminar, February 4, 1986*, ed. Martin E. Alexander (study NOR-5-05 [NOR-5-19], file report no. 15) (Canadian Forestry Service, Edmonton, 1986), 20–30, http://cfs.nrcan.gc.ca/pubwarehouse/pdfs/24290.pdf.
25   Ibid.
26   Ibid.
27   H. Wexler, "The Great Smoke Pall of September 24–30, 1950," *Weatherwise*, December 1950, 129–34, 142.
28   Smith, "The Widespread Smoke Layer," 180–84.
29   Wexler, "The Great Smoke Pall," 129–34, 142.
30   Ibid.
31   Smith, "The Widespread Smoke Layer," 182.
32   Ibid.
33   Elsley, "Alberta Forest-Fire Smoke," 22–26.
34   *Toronto Star*, "Alberta Smoke Covers Toronto," September 25, 1950.
35   Ibid.
36   Smith, "The Widespread Smoke Layer," 180–84.
37   Wexler, "The Great Smoke Pall," 131.
38   Smith, "The Widespread Smoke Layer," 183. Millibars (mb) are a unit of atmospheric pressure.
39   Ibid.
40   Ibid.
41   Ibid.
42   *Edmonton Bulletin*, "100 Bush Fires Ravaging North," September 25, 1950.
43   Wexler, "The Great Smoke Pall," 129.
44   Smith, "The Widespread Smoke Layer," 182.
45   Wexler, "The Great Smoke Pall."
46   *Toronto Star*, "Blue Moon, Blue Sun, Too Give U.K., Europe Jitters, Alberta Smoke Is Blamed," September 27, 1950; Smith, "The Widespread Smoke Layer," 180.

THREE   DARK DAYS IN THE PAST

1   F.G. Plummer, "Forest Fires: Their Causes, Extent and Effects, with a Summary of Recorded Destruction and Loss" (USDA Forest Service bulletin no. 117, Washington, DC, 1912).
2   *Wikipedia*, s.v. "Plagues of Egypt," accessed July 24, 2011, http://en.wikipedia.org/wiki/Plagues_of_Egypt.

3 *Wikipedia*, s.v. "Crucifixion Darkness and Eclipse," accessed July 24, 2011, http://en.wikipedia.org/wiki/Crucifixion_darkness_and_eclipse.

4 Erin R. McMurry et al., "Fire Scars Reveal Source of New England's 1780 Dark Day," *International Journal of Wildland Fire* 16 (2007): 266–70.

5 Plummer, "Forest Fires."

6 Ibid.

7 Henry Youle Hind, *Explorations in the Interior of the Labrador Peninsula: The Country of the Montagnais and Nasquapee Indians*, vol. 1 (London: Longman, Green, Longman, Roberts, and Green, 1863).

8 Ibid.

9 M.E. Alexander, "'Lest We Forget': Canada's Major Wildland Fire Disasters of the Past, 1825–1938," in *Proceedings of 3rd Fire Behavior and Fuels Conference, October 25–29, 2010, Spokane, Washington, US* (Birmingham, AL: International Association of Wildland Fire, 2010).

10 Ibid.; Government of New Brunswick, *Great Miramichi Fire* (Fredericton: New Brunswick Natural Resources, 2009), http://web.archive.org/web/20101013094259/http://www.gnb.ca/0079/miramichi_fire-e.asp.

11 Timothy Egan, *The Big Burn: Teddy Roosevelt and the Fire That Saved America* (Boston: Houghton Mifflin Harcourt, 2009).

12 Ibid.

13 C.F. Talman, "Dark Days and Forest Fires," *Scientific American*, March 6, 1915, 229–30.

14 Stephen J. Pyne, *Awful Splendor: A Fire History of Canada* (Vancouver: University of British Columbia Press, 2007).

15 Amanda Dawn Annand, "The 1910 Fires in Alberta's Foothill and Rocky Mountain Regions" (Hon. BSc. thesis, University of Victoria, Victoria, BC, 2010).

16 Rick Arthur, personal communication with the author, February 3, 2013. Arthur mentioned that fires also occurred outside the Forest Reserves in 1910.

17 Jim Petersen "The 1910 Fire," *Evergreen Magazine*, Winter 1994–95, http://www.idahoforests.org/fires.htm.

18 USDA Forest Service, Idaho Panhandle National Forests, *When the Mountains Roared: Stories of the 1910 Fire* (Coeur d'Alene, ID: USDA Forest Service, Idaho Panhandle National Forests, 1978).

19 Fred McClement, *The Flaming Forests* (Toronto: McClelland and Stewart, 1969).

20 Ibid.

21 The 75th Anniversary of the Great Fire of 1922 Committee, *The Great Fire of 1922* (Haileybury, ON: Haileybury Heritage Museum, 1999, repr. May 2000).

22 Ibid.

23 L.E. Johnson, "Commemorating the 80th Anniversary of the Fires of 1918," *Minnesota History Interpreter* 26, no. 6 (1998).

24 H. Lyman, "Smoke from Minnesota Forest Fires," *Monthly Weather Review* 46, no. 11 (1919): 506–09.

25 Wexler, "The Great Smoke Pall," 142.

26 George C. Joy, "The Conflagration: Western Possibilities of Sweeping Fires Like Those of Minnesota and Canada" (address delivered at Annual Conference, Western Forestry and Conservation Association, Portland, November 25, 1922) (Portland: Oregon State Board of Forestry, 1922), 4.

27 Ibid.

28 William G. Morris, "Forest Fires in Western Oregon and Western Washington," *Oregon Historical Quarterly* 35 (1934): 313–39.

29 Joy, "The Conflagration," 4.

30 Morris, "Forest Fires in Western Oregon and Western Washington."

31 "The Great Peshtigo Fire of 1871," accessed November 22, 2009, http://www.peshtigofire.info/.

32 "What Were the Largest Fires (Lower 48 States)?" *Wildlandfire.com*, accessed April 2, 2009, http://www.wildlandfire.com/docs/imwtk-lg-fires.htm.

33 Ibid.

34 R. McClure, "Washington's 'Awful Conflagration'—The Yacolt Fire of 1902," *Fire Management Today* 65, no. 1 (2005): 24–27.

35 Ibid.

36 Joy, "The Conflagration," 4.

37 R.J. Roder, "The Great Fire of 1909," *Western Producer*, October 29, 1964, 17, 19.

38 *Morning Bulletin* (Edmonton), "Lac La Biche Village in Ashes; Entire District is Homeless; Condition of People Perilous," May 21, 1919.

39 Ibid.

40 Government of Canada, *Review of Fire Season 1919 on Dominion Forest Reserves & Fire Ranging Districts* (Ottawa: Dominion Forestry Branch, 1919).

41 P.J. Murphy, notes on a telephone conversation with Mike Maccagno, March 28, 1984, accession no. PR2008.0109/12, 2 p., Provincial Archives of Alberta, Edmonton.

42 Government of Canada, *Report of the Director of Forestry For The Fiscal Ended March 31, 1920* (Ottawa: Department of the Interior, 1921), 5, http://cfs.nrcan.gc.ca/bookstore_pdfs/30885.pdf.

43 Joseph F. Dion, *My Tribe, the Crees* (Calgary, AB: Glenbow Museum, Calgary, 1979).

44 P.J. Murphy, notes on a telephone conversation with Mike Maccagno, March 28, 1984, accession no. PR2008.0109/12, Provincial Archives of Alberta, Edmonton.

45 Plummer, "Forest Fires."

46 R. Seitz, "Siberian Fire as 'Nuclear Winter' Guide," *Nature* 323 (September 11, 1986): 116–17.

47 Area estimated from Figure 1 in Seitz, "Siberian Fire as 'Nuclear Winter' Guide," 116–17.

48 Seitz, "Siberian Fire as 'Nuclear Winter' Guide," 116–17.

49 Ibid.

50 M.D.F. Udvardy, *Dynamic Zoogeography, with Special Reference to Land Animals* (New York: Van Nostrand Reinhold Co., 1969).

51 Harrison E. Salisbury, *The Great Black Dragon Fire: A Chinese Inferno* (Boston: Little Brown and Company, 1989).

52 Joy, "The Conflagration," 4.

53 Simon Winchester, *Krakatoa: The Day the World Exploded: August 27, 1883* (New York: HarperCollins, 2003).

54 George Pararas-Carayannis, "The Great Explosion of the Krakatau Volcano ('Krakatoa') of August 26, 1883, in Indonesia," *The Tsunami Page of Dr. George P.C.*, accessed July 24, 2011, http://www.drgeorgepc.com/Volcano1883Krakatoa.html.

55 Ibid.

56 Ibid.

57 Ibid.

58   Ibid.

59   United States Geological Survey, "The Cataclysmic 1991 Eruption of Mount Pinatubo, Philippines" (US Geological Survey fact sheet 113-97), last modified February 28, 2005, http://pubs/usgs.gov/fs/1997/fs113-97/.

60   Wexler, "The Great Smoke Pall," 132.

61   Wexler, "The Great Smoke Pall," 129.

FOUR   BLUE MOON, BLUE SUN

1   Bull, "Blue Sun and Moon," 167–69.

2   Ibid.

3   Ibid.

4   *Times* (London), "Blue Tint to the Sun: Haze at 30,000 Feet," September 27, 1950.

5   Alan Watson, "Moon Blues" (post), "Physics Questions/Problems 12/97," "Once in a Blue Moon," "Lastword," (Yacov Kantor's personal website), accessed September 24, 2014, http://star.tau.ac.il/QUIZ//97/lastword.html.

6   Bull, "Blue Sun and Moon," 167–69.

7   Ibid.

8   Ibid.

9   CBC-Radio, *Round Up.*

10   Ibid.

11   Ibid.

12   *Arbroath Guide*, "Rare Phenomena: Blue Sun and Blue Moon Observed throughout Scotland," September 30, 1950.

13   Ibid.

14   Wexler, "The Great Smoke Pall," 129–35.

15   Ibid.

16   Bull, "Blue Sun and Moon," 167–69.

17   Bradford Landmark Society, "The Day the Sun Went out in Bradford. The Blackout— September 24, 1950," Bradford Landmark Society (Bradford, PA), accessed October 14, 2014, http://www.bradfordlandmark.org/index.php?The%20Day%20the%20Sun%20 Went%20Out%20In%20Bradford.

18   Wexler, "The Great Smoke Pall," 129–35.

19   Ibid.

20   Ibid.

21   Jerome Barlowe and William Roye et al., *Rede Me and Be Nott Wrothe, for I Saye No Thinge But Trothe*, ed. D.H. Parker (Toronto: University of Toronto Press, 1992), originally published in 1528.

22   Philip Hiscock, "Folklore of the 'Blue Moon,'" *International Planetarium Society*, 1999, accessed September 20, 2014, http://www.ips-planetarium. org/?page=a_hiscock1999.

23   Ibid.

24   Ibid.

25   Ibid.

26   Ibid.

27   Ibid.

28   Ibid.

29   James Pruett, "Once in a Blue Moon," *Sky & Telescope*, March 1946.

30   Laurence Lafleur, "Sauce for the Gander," (question and answer column), *Sky & Telescope*, July 1943.

31   Roger Sinnott, Donald Olson, and Richard Fienberg, "What's a Blue Moon?" *Sky & Telescope*, May 1999.

32   Rudolf Penndorf, "On the Phenomenon of the Colored Sun, Especially the 'Blue' Sun of September 1950" (geophysical research papers no. 20, Geophysics Research Directorate, Air Force Cambridge Research Center, Cambridge, MA, 1953).

33   Hendrik Christoffel Van de Hulst, *Light Scattering by Small Particles* (New York: Dover Publications, 1981), originally published by John Wiley & Sons, 1957.

34   Penndorf, "On the Phenomenon of the Colored Sun."

35   Craig F. Bohren and Donald R. Huffman, *Absorption and Scattering of Light by Small Particles* (New York: John Wiley & Sons, 1983).

36   Robert Wilson, "The Blue Sun of 1950 September," *Monthly Notices of the Royal Astronomical Society* 111 (1951): 478.

37   Roger D. Ottmar, "Smoke Source Characteristics," in *Smoke Management Guide for Prescribed and Wildland Fire, 2001 Edition*, eds. Colin C. Hardy et al. (rep. PMS 420-2/NFES 1279, National Wildfire Coordination Group, 2001), 89–106.

38   D.E. Ward, "Smoke from Wildland Fires," in *Health Guidelines for Vegetation Fire Events, Lima, Peru, October 6–9, 1998, Background Papers*, eds. Kee-Tai Goh et al. (N.p.: World Health Organization, 1999), 70–85.

39   N.J. Payne et al., "Combustion Aerosol from Experimental Crown Fires in a Boreal Forest Jack Pine Stand," *Canadian Journal of Forest Research* 34 (2004): 1627–33.

40   W.M. Porch, C.S. Atherton, and J.E. Penner, "Atmospheric Optical Effects of Aerosols in Large Fires," in *Proceedings Ninth Conference on Fire and Forest Meteorology, April 21–24, 1987, San Diego, CA* (Boston: American Meteorological Society, 1987), 253–57.

41   Ibid.

42   P. Pesic, "A Simple Explanation of Blue Suns and Moons," *European Journal of Physics* 29 (2008): N31–N36.

43   Porch, Atherton, and Penner, "Atmospheric Optical Effects," 253–57.

44   W.M. Porch, "Blue Moons and Large Fires," *Applied Optics* 28, no. 10 (1989): 1778–84.

45   Craig F. Bohren, *Clouds in a Glass of Beer: Simple Experiments in Atmospheric Physics* (Mineola, NY: Dover Publications, 1987).

46   Ibid.

47   Planofworld, "Rare Blue Sun Phenomenon Photographed near Giza Pyramids, December 14, 2006," *The World Of...* (blog), September 14, 2009, http://planofworld. blogspot.com/2009/09/rare-blue-sun-phenomenon-photographed.html.

48   Aden Meinel, *Sunsets, Twilights, and Evening Skies* (New York: Cambridge University Press, 1983).

49   Wexler, "The Great Smoke Pall," 134–35.

1   Mary Hardin and Ralph Kahn, "Aerosols and Climate Change," *Earth Observatory*, last updated November 1, 2010, http://earthobservatory.nasa.gov/Features/Aerosols/what_are_aerosols_1999.pdf.

2   N.A. Marley et al., "Residence Times of Fine Tropospheric Aerosols as Determined by ²¹⁰Pb Progeny," in *Symposium on Atmospheric Chemistry Issues in the 21st Century* (Boston: American Meteorological Society, 2000), 1–5.

3   TSI Incorporated, "Health-Based Particle-Size-Selective Sampling" (application note ITI-050, 2009), http://www.tsi.com/uploadedFiles/Product_Information/Literature/Application_Notes/ITI-050.pdf.

4   K. Yalcin et al., "Ice Core Paleovolcanic Records from the St. Elias Mountains, Yukon, Canada," *Journal of Geophysical Research* 112 (2007): D08102, doi:10.1029/2006JD007497.

5   M.D. Fromm et al., "Observations of Boreal Forest Fire Smoke in the Stratosphere by POAM III, SAGE II and Lidar in 1998," *Geophysical Research Letters* 27 (2000): 1407–10.

6   M.D. Fromm and R. Servranckx, "Transport of Forest Fire Smoke above the Troposphere by Supercell Convection," *Geophysical Research Letters* 30 (2002): 1542–45.

7   G. Holdsworth, "Evidence for a Link between Atmospheric Thermonuclear Detonations and Nitric Acid," *Nature* 324 (1986): 551–53.

8   Dr. Gerald Holdsworth (University of Calgary), personal communication (email) with the author, December 2011.

9   Ottmar, "Smoke Source Characteristics," 89–106.

10 H. Levy et al., "Simulated Tropospheric NO_x: Its Evaluation, Global Distribution and Individual Source Contributions," *Journal of Geophysical Research-Atmospheres* 104 (1999): 26279–306, doi:10.1029/1999JD900442.

11 Dr. Meredith Hastings (Brown University), personal communication (email) with the author, December 11, 2011.

12 FRAMES Resource Cataloguing System, Alaska Fire History Database, accessed October 19, 2014, https://www.frames.gov/rcs/10000/10436.html.

13 *Edmonton Journal*, "Smoke Blanket from North Fires," September 15, 1950.

14 E. Kalnay et al., "The NCEP/NCAR 40-Year Reanalysis Project," *Bulletin of the American Meteorological Society* 77 (1996): 437–71.

15 Zeke Hausfather, "The Water Vapor Feedback," *Yale Climate Connections*, February 4, 2008, http://www.yaleclimateconnections.org/2008/02/common-climate-misconceptions-the-water-vapor-feedback-2/.

16 R. Sausen et al., "A Diagnostic Study of the Global Distribution of Contrails. Part I: Present Day Climate," *Theoretical and Applied Climatology* 61 (1998): 127–41.

17 D.J. Travis, A.M. Carleton, and R.G. Lauritsen, "Contrails Reduce Daily Temperature Range," *Nature* 418, no. 601 (2002): doi:10.1038/418601a.

18 R.P. Turco et al., "Nuclear Winter: Global Consequences of Multiple Nuclear Explosions," *Science* 222 (1983): 4630.

19 "Japanese Balloon Bomb Attack on the US," accessed October 22, 2011, http://www.bookmice.net/darkchilde/japan/balloon.html.

20 H.M. Schmeck, "Climate Changes Endanger World's Food Output," *New York Times*, August 8, 1974, http://www.wmconnolley.org.uk/sci/iceage/nytimes-1974-08-08.pdf.

21 P. Gwynne, "The Cooling World," *Newsweek*, April 28, 1975, http://www.denisdutton.com/cooling_world.htm.

22 Ibid.

23 Ibid.

24 D. McRae, "Prescribed Burning for Stand Conversion in Budworm-Killed Balsam Fir: An Ontario Case History," *Forestry Chronicle* 62, no. 2 (1986): 96–100.

25 Ibid.

26 Amber J. Soja et al., "ARTAS: The Perfect Smoke," *Canadian Smoke Newsletter*, Fall 2008, 2–7, http://www.sk.lung.ca/canadian_smoke_newsletters/fall_2008.pdf.

27 Letter from Craig Chandler (retired director of fire research, USDA Forest Service) to Peter Murphy (associate dean, Forestry, Faculty of Agriculture and Forestry, University of Alberta, Edmonton), July 20, 1984, letter in author's possession.

28 Ibid.

29 P.J. Crutzen and J.W. Birks, "The Atmosphere after Nuclear War: Twilight at Noon," in *The Aftermath: The Human and Ecological Consequences of Nuclear War*, ed. J. Peterson, special issue of *Ambio* 11, no. 2–3 (1982): 76–83.

30 Ibid.

31 Lawrence Badash, *A Nuclear Winter's Tale: Science and Politics in the 1980s* (Cambridge, MA: MIT Press, 2009.

32 Sarah Henderson and Michael Brauer, "Fire Smoke and Human Health," *Canadian Smoke Newsletter*, Spring 2008, 6–7, http://www.sk.lung.ca/canadian_smoke_newsletters/spring_2008.pdf.

33 Ibid.

34 Leonard F. DeBano, Daniel G. Neary, and Peter F. Ffolliott, *Fire's Effects on Ecosystems* (New York: John Wiley & Sons, 1998).

35 Harriet Ammann, *Wildfire Smoke: A Guide for Public Health Officials* (Sacramento, CA: Office of Environmental Health Hazard Assessment, 2001).

36 Henry Antoine Des Voeux, "Fog and Smoke" (paper presented at the Royal Institute of Public Health, London Congress Meeting, July 1905, London, UK). The term *smog* was used by the *Daily Graphic*, a London newspaper, on July 26, 1905.

37 United States Environmental Protection Agency, *Smog—Who Does It Hurt? What You Need to Know about Ozone and Your Health* (Washington, DC: United States Environmental Protection Agency, 1999), http://epa.gov/air/ozonepollution/pdfs/smog.pdf.

38 Ibid.

39 Ontario Medical Association, *The Illness Costs of Air Pollution: 2005–2026 Health and Economic Damage Estimates* (Toronto: Ontario Medical Association, June 2005), https://www.oma.org/Resources/Documents/e2005HealthAndEconomicDamageEstimates.pdf.

40 Ibid.

41 Ibid.

42 Canadian Medical Association, *No Breathing Room: National Illness Costs of Air Pollution. NICAP Summary Report* (Ottawa: Canadian Medical Association, August 2008).

43 Ibid.

44 United States Environmental Protection Agency, *Air Quality Index*, accessed October 2, 2011, http://www.epa.gov/airnow/smoke2/smoke2.html.

45 City of Toronto, "Smog Alert Days 2005," accessed July 13, 2005, http://www.city.toronto.on.ca/health/smog/smog_new.htm.

46   BBC News, "The Great Smog of London," December 5, 2002, http://news.bbc.
      co.uk/2/hi/uk_news/england/2545759.stm.
47   London Air Quality Network (LAQN), "LAQN Monitoring Sites," accessed September
      24, 2014, http://www.londonair.org.uk/london/asp/publicdetails.asp?region=0.
48   T. Spears, "Blackout Sheds Light on Pollution," Edmonton Journal, May 31, 2004.
49   B.D. Amiro et al., "Direct Carbon Emissions from Canadian Forest Fires, 1959–1999,"
      Canadian Journal of Forest Research 31 (2001): 512–25.
50   Canadian Medical Association, No Breathing Room.
51   Johnston, "Estimated Global Mortality Attributable to Smoke from Landscape Fires."
52   R. Rittmaster et al., "Economic Analysis of Health Effects from Forest Fires,"
      Canadian Journal of Forest Research 36 (2006): 868–77.
53   Ibid., R. Rittmaster et al., "Erratum: Economic Analysis of Health Effects from
      Forest Fires," Canadian Journal of Forest Research 38 (2008): 908; Chisholm Fire Review
      Committee, Chisholm Fire Review Committee Final Report, Submitted to the Minister of Alberta
      Sustainable Resource Development, October 2001 (Edmonton: Alberta Environment and
      Sustainable Resource Development, 2001). The Chisholm Fire Review Committee
      2001 report states the firefighting cost was $10 million. This is an error. Actual
      firefighting cost was $30 million. The estimated health cost, however, was approxi-
      mately equal to the loss of homes, buildings, and infrastructure.
54   United Nations Environment Programme, and Centre for Clouds, Chemistry,
      and Climate (C4), The Asian Brown Cloud: Climate and Other Environmental Impacts. UNEP
      Assessment Report (Pathumthani, Thailand: UNEP and C4, 2002), http://www.rrcap.
      unep.org/abc/impactstudy/.
55   V.M. Ramanathan et al., Atmospheric Brown Clouds: Regional Assessment Report with Focus
      on Asia (Nairobi, KE: United Nations Environment Programme, 2008), http://www.
      unep.org/pdf/ABCSummaryFinal.pdf.
56   World Health Organization, "Tackling the Global Clean Air Challenge," news
      release, September 26, 2011), http://www.who.int/mediacentre/news/
      releases/2011/air_pollution_20110926/en/index.html.
57   Ramanathan et al., Atmospheric Brown Clouds.
58   Michael Kovrig, "Quebec Forest Fires Blanket Northeastern United States eith
      Smoke," Seacoastonline.com, July 8, 2002, http://www.seacoastonline.com/
      articles/20020708-NEWS-307089993?cid=sitesearch.
59   Stan Coloff, personal communication with author, Fire Research Coordination
      Council Meeting, July 9, 2002, Washington, DC.
60   Michael Kovrig, "Quebec Fires Cast Pall over Northeast United
      States," post-gazette.com, July 8, 2002, http://www.post-gazette.com/
      nation/20020708quebec0708p2.asp.
61   Joint Action Implementation Coordinating Committee, Report to the Council of Ministers
      of Environment: An Update in Support of the Canada-Wide Standards for Particulate Matter and
      Ozone (N.p.: Joint Action Implementation Coordinating Committee, 2005), http://
      www.ccme.ca/files/Resources/air/pm_ozone/pm_03_update_2005_e.pdf.
62   Ontario Ministry of the Environment, Air Quality in Ontario 2002 Report (Toronto:
      Ontario Ministry of the Environment, Environmental Monitoring and
      Reporting Branch, 2002), http://www.airqualityontario.com/downloads/
      AirQualityInOntarioReport2002.pdf.

63   D. Lavoué, S.L. Gong, and B.J. Stocks, "Modelling Emissions from Canadian Wildfires: A Case Study of the 2002 Quebec Forest Fires," *International Journal of Wildland Fire* 16, no. 6 (2007): 649–63.

64   A. Sapkota et al., "Impact of the 2002 Canadian Forest Fires on Particulate Matter Air Quality in Baltimore City," *Environmental Science and Technology* 39 (2005): 24–32.

65   David Lavoué, "Quebec Wildfires Spread Smoke to Eastern Canada and New England," *Canadian Smoke Newsletter*, Fall/Winter 2010, 15–17, http://www.sk.lung. ca/canadian_smoke_newsletters/fall_winter_2010.pdf; NASA, "Canadian Fires Send Smoke over New England," *Earth Observatory*, June 2, 2010, http://earthobservatory. nasa.gov/IOTD/view.php?id=44138&src=eoa-iotd.

66   D.E. Henderson, "Prescribed Burns and Wildfires in Colorado: Impacts of Mitigation Measures on Indoor Air Particulate Matter," *Journal of the Air and Waste Management Association* 55, no. 10 (2005): 1516–26.

67   S.E. Page et al., "Carbon Released during Peatland Fires in Central Lalimantan, Indonesia in 1997," *Nature* 420 (2002): 61–65. One gigatonne (Gt) is one billion tonnes.

68   BBC News, "Forest Fire Smoke Blankets Moscow," July 31, 2002, http://news.bbc. co.uk/1/hi/world/europe/2163390.stm.

69   Q. Schiermeier, "Climate Change: That Sinking Feeling," *Nature* 435 (2005): 732–33, doi:10.1038/435732a.

70   M.R. Turetsky, W. Donahue, and B.W. Benscoter, "Experimental Drying Intensifies Burning and Carbon Losses in a Northern Peatland," *Nature Communications* (November 2011), doi:10.1038/ncomms1523.

71   Kamonayi Mubenga, "The Alaska, Yukon and British Columbia Smoke across the U.S.," *U.S. Air Quality, The Smog Blog*, January 24, 2005, http://alg.umbc.edu/usaq/ archives/000844.html.

72   NASA, "Smoke Signals from the Alaska and Yukon Fires, July 22, 2004," *Earth Observatory*, http://earthobservatory.nasa.gov/IOTD/view.php?id=4676.

73   D. Lavoué and B.J. Stocks, "Emissions of Air Pollutants by Canadian Wildfires from 2000 to 2004," *International Journal of Wildland Fire* 20, no. 1 (2011): 17–34, doi:10.1071/WF08114.

74   CBC News, "700 Leave Homes as Ontario Forest Fires Burn," September 13, 2006, http://www.cbc.ca/news/canada/story/2006/09/13/ontario-fires.html.

75   Forestry Canada Fire Danger Group, *Development and Structure of the Canadian Forest Fire Behaviour Prediction System* (information report ST-X-3, Ottawa: Forestry Canada, 1992).

76   Lavoué and Stocks, "Emissions of Air Pollutants by Canadian Wildfires."

77   W.J. de Groot et al., "Estimating Direct Carbon Emissions from Canadian Wildland Fires," *International Journal of Wildland Fire* 16 (2007): 593–606; W.A. Kurz and M.J. Apps, "Developing Canada's National Forest Carbon Fluxes in the Canadian Forest Sector," *Ecological Applications* 9 (2006): 526–47, doi:10.1007/S11027-006-1006-6.

78   Amiro et al., "Direct Carbon Emissions from Canadian Forest Fires."

SIX   THE BIG WIND

1   Murphy and Tymstra, "The 1950 Chinchaga River Fire."
2   Ibid.

3   Ibid.
4   LaFoy, "The 1950 Chinchaga River Fire."
5   Murphy and Tymstra, "The 1950 Chinchaga River Fire."
6   C.E. Van Wagner, *Development and Structure of the Canadian Forest Fire Weather Index System* (Forestry technical report 35, Ottawa: Canadian Forestry Service, 1987).
7   Cordy Tymstra, "Fire Behaviour Study—Chinchaga Fire," program output files, FOR 451: Intermediate Forestry Problems course, University of Alberta, 1985.
8   Forestry Canada Fire Danger Group, *Development and Structure of the Canadian Forest Fire Behaviour Prediction System*.
9   Cordy Tymstra, "Fire Behaviour Study—Chinchaga Fire," program output files, FOR 451: Intermediate Forestry Problems course, University of Alberta, 1985.
10 Smith, "The Widespread Smoke Layer," 180–84.
11 Department of Transport, Meteorological Division, *Monthly Record of Meteorological Observations in Canada* (September 1950).
12 Environment Canada, *Canadian Normals: Precipitation 1941–1970*, vol. 2 (Downsview, ON: Atmospheric Environment, 1973); Department of Transport, Meteorological Division, *Monthly Record of Meteorological Observations in Canada* (August 1950).
13 Department of Transport, Meterological Division, *Monthly Record of Meteorological Observations in Canada* (September 1950).
14 Environment Canada, *Canadian Normals*; Department of Transport, Meteorological Division, *Monthly Record of Meteorological Observations in Canada* (August 1950).
15 Alberta Department of Lands and Forests, *Rat Fire Report*.
16 British Columbia Department of Lands and Forests, *Whisp Fire Report*.
17 Ibid.
18 Murphy and Tymstra, "The 1950 Chinchaga River Fire."
19 Alberta Department of Lands and Forests, *Outlaw Fire Report*.
20 Atmospheric Environment Service, meteorological observations on microfilm, obtained from Climatological Services Division, Atmospheric Environment Service, Downsview, ON, 1986.
21 Murphy and Tymstra, "The 1950 Chinchaga River Fire."
22 Memo from B.H. Johnson (forest land use officer, Alberta Forest Service, Peace River) to Lou Foley (forest protection officer, Alberta Forest Service, Peace River), January 10, 1983.
23 B. Janz and N. Nimchuk, "The 500 mb Anomaly Chart—A Useful Fire Management Tool," in *Proceedings of the Eighth Conference on Fire and Forest Meteorology, April 29–May 2, 1985, Detroit, MI* (Bethesda, MD: Society of American Foresters, 1985), 233–338.
24 R.W. Thistlethwaite, "Survey of Boundary Northward from the Termination of the 1923 Survey," in *Report and Field Journal, Summer Season 1950 and Winter Season 1950–1951* (Ottawa: Department of the Interior, 1952), Glenbow Archives, Calgary; R.W. Thistlethwaite, "Alberta–British Columbia Boundary: Positions and Descriptions of Monuments Established on Alberta–British Columbia Boundary," in *Monument Book* (Ottawa: Department of the Interior, 1951), Glenbow Archives, Calgary.
25 R.W. Thistlethwaite, "Survey of Boundary Northward," 70.
26 Ibid., 72.
27 Ibid., 77.
28 Ibid.
29 Ibid., 29.

30 *Edmonton Journal*, "Rain Relieves Fire Threat over Wide Alberta Section," September 26, 1950.

SEVEN POLICY CHANGES

1 E.S. Huestis, *Report of the Director of Forestry. Annual Report of the Department of Lands and Forests of the Province of Alberta for the Fiscal Year Ended March 31st 1950* (Edmonton: King's Printer, 1951), 26–72.

2 Ibid.

3 Ibid.

4 Ibid.

5 Ibid.

6 Government of Canada, *Forest Fire Losses in Canada 1949* (Ottawa: Forestry Branch, Forest Research Division, Department of Resources and Development, 1950).

7 Huestis, *Report of the Director of Forestry ... for the Fiscal Year Ended March 31st 1950*, 26–72.

8 Ibid.

9 Ibid.

10 Government of Canada, *Report of the Department of Resources and Development for the Fiscal Year Ended March 31, 1950* (Ottawa: Forestry Branch, 1950), 103–42.

11 Huestis, *Report of the Director of Forestry ... for the Fiscal Year Ended March 31st 1950*.

12 Ibid.

13 *Regulations for the prevention of forest and prairie fires*, OIC 1119/50, (1950) A Gaz, 1331–1332, (*The Forests Act*).

14 Ibid.

15 Ibid.

16 *Battle River Herald*, "Forest Fires Are Costly," July 27, 1950.

17 *Peace River Record-Gazette*, "Forestry Car Dates Announced for District Visit," June 15, 1950.

18 J.A. Doyle, "A Survey of Forest Fire Causes and Suggested Corrective Methods," *Forestry Chronicle* 27, no. 4 (1951): 335–48.

19 Huestis, *Report of the Director of Forestry ... for the Fiscal Year Ended March 31st 1949*, 18–55.

20 Huestis, *Report of the Director of Forestry ... for the Fiscal Year Ended March 31st 1950*, 26–72.

21 E.S. Huestis, *Report of the Director of Forestry. Annual Report of the Department of Lands and Mines of the Province of Alberta for the Fiscal Year Ended March 31st 1951* (Edmonton: King's Printer, 1952), 23–45.

22 *Peace River Record-Gazette*, "Nature and People Combined to Cut Forest Fire Losses," August 17, 1950.

23 Doyle, "A Survey of Forest Fire Causes," 335–48.

24 Blefgen, *Report of the Director of Forestry ... for the Fiscal Year Ended March 31st 1943*, 34–55.

25 Ibid.

26 S. Janzen, "The Burning North: A History of Fire and Fire Protection in the NWT" (MA thesis, University of Alberta, 1990).

27 Doyle, "A Survey of Forest Fire Causes," 335–48.

28   *Times Colonist* (Victoria), "Fire Forces Tumble Ridge to Evacuate," *Canada. com*, July 4, 2006, http://www.canada.com/victoriatimescolonist/story. html?id=30946ab8-1046-4b2f-9961-0ffe0681158e&k=51191.

29   D.A. MacGibbon, *The Canadian Grain Trade 1931–1951* (Toronto: University of Toronto Press, 1952).

30   Leonard, *The Last Great West*.

31   *Certain provincial lands, including school lands, withdrawn from settlement*, 01c 113/48, (1948) A Gaz, 182 (*The Provincial Lands Act*).

32   Gary Filmon and Firestorm 2003 Provincial Review Team, *Firestorm 2003 Provincial Review* (Vancouver: Government of British Columbia, 2004), http://bcwildfire.ca/History/ ReportsandReviews/2003/FirestormReport.pdf.

33   Ibid.

34   Ibid.

35   N. Moharib and T. White, "BC City Battles Wildfires: About 11,000 Residents Flee," *Toronto Sun*, July 20, 2009, http://www.torontosun.com/news/ canada/2009/07/19/10188246.html.

36   Huestis, *Report of the Director of Forestry … for the Fiscal Year Ended March 31st 1949*, 18–55.

37   Alberta Forest Products Association, *Alberta's Forests: A Renewable Story* (Edmonton: Alberta Forest Productions Association, n.d.), accessed September 24, 2014, https:// www.albertaforestproducts.ca/sites/default/files/downloads/Renewable%20 Story.pdf.

38   Nordic Forest Owners' Association, "Forests in Sweden," *Nordic Family Forestry*, accessed March 24, 2011, http://www.nordicforestry.org/facts/Sweden.asp.

39   Huestis, *Report of the Director of Forestry … for the Fiscal Year Ended March 31st 1951*, 23–45.

40   Ibid.

41   T.F. Blefgen, *Report of the Director of Forestry. Annual Report of the Department of Lands and Mines of the Province of Alberta for the Fiscal Year Ended March 31st 1945* (Edmonton: King's Printer, 1946), 34–55.

42   Government of British Columbia, *Report of Forest Service* (Victoria: Department of Lands and Forests, 1951).

43   Ibid.

44   Ibid.

45   Ibid.

46   McKee, "Fire Protection in British Columbia," 94–96.

47   Government of British Columbia, *Report of Forest Service*.

48   Ibid.

49   Ibid.

50   Memorandum from Eric Huestis (acting assistant director of forestry) to Deputy Minister J. Harvie, March 29, 1945, Alberta Government Library, Edmonton. The memorandum states, "In 1940 the Government laid down a policy whereby it was decided that there would be no expenditure of funds for the fighting of fire north of a specified line in the northern part of the Province except where such fires were endangering lives or threatening property."

51   Memorandum from Eric Huestis (acting assistant director of forestry) to Deputy Minister J. Harvie, March 29, 1945, Alberta Government Library, Edmonton.

52 Memorandum from Frank Neilson (chief timber inspector) to Acting Assistant Director of Forestry Eric Huestis, September 27, 1945, Alberta Government Library, Edmonton.

53 Ibid.

54 Platt, "Fire Suppression."

55 E.S. Huestis, "Early Days in Alberta Forestry" (talk given to new Alberta Forest Service forest officers during an orientation course at the Forest Technology School in Hinton, AB, January 31, 1972), transcribed and ed. P.J. Murphy, July 1986.

56 E.S. Huestis. *Report of the Director of Forestry. Annual Report of the Department of Lands and Mines of the Province of Alberta for the Fiscal Year Ended March 31st 1952* (Edmonton: King's Printer, 1953), 23–45.

57 Canadian Institute of Forestry, *Forest Fire Protection in Alberta: A Review and Recommendations* (N.p.: Canadian Institute of Forestry, Rocky Mountain Section, 1954).

58 LaFoy, "The 1950 Chinchaga River Fire."

59 Ibid.

60 British Columbia Department of Lands and Forests, *Want Lake Fire Report* (Prince George, BC: Forest Service, 1950).

61 CBC, "A Wealth of Oil," *Canada: A People's History*, accessed October 22, 2011, http://www.cbc.ca/history/EPISCONTENTSE1EP15CH3PA2LE.html.

62 Huestis, *Report of the Director of Forestry … for the Fiscal Year Ended March 31st 1950*, 26–72.

63 Government of Alberta, *Forest Cover Map of Alberta* (Edmonton: Department of Lands and Forests, 1957).

64 Government of Alberta, *Map of Areas (1,000 Acres and over) Devastated by Forest Fires 1938 to 1949 Inclusive* (Edmonton: Department of Lands and Forests, 1949).

65 J.A. Doucet, *Timber Conditions in the Smoky River Valley and Grande-Prairie Country* (Forestry Branch bulletin no. 53, Department of the Interior, Ottawa, 1915).

66 Government of Alberta, *Alberta Forest Inventory* (Edmonton: Department of Lands and Forests, Forest Service, 1968).

67 Alberta Department of Lands and Forests, *Fire 28A-2 Report* (Peace River: Forest Service, 1958).

68 Jack Grant, interview by Peter J. Murphy (January 26, 1999), in *Alberta Forest Service History Project: 2005 – 75th Anniversary* (Edmonton: Alberta Land and Forest Service, 1999).

69 Ibid.

70 Ibid., 3.

71 Gordon Reid, *Around the Lower Peace* (Peace River: Lower Peace Publishing Company, 1978), 5–8.

72 P.J. Murphy, R.E. Stevenson, D. Quintilio, and S. Ferdinand, *The Alberta Forest Service, 1930–2005: Protection and Management of Alberta's Forests* (Edmonton: Alberta Sustainable Resource Development, Government of Alberta, 2006).

73 Government of Alberta, *Fallen Firefighter Memorial Dedication Ceremony*, September 20, 2008, (Hinton, AB: Hinton Training Centre), DVD.

1   Weyerhaeuser Canada, "The Ed Olney Story: Part I," *The Lakeside Leader* (Slave Lake, AB), February 5, 1997.

2   E.S. Huestis, *Report of the Director of Forestry. Annual Report of the Department of Lands and Forests of the Province of Alberta for the Fiscal Year Ended March 31st 1961* (Edmonton: Queen's Printer, 1962), 28–75.

3   Weyerhaeuser Canada, "The Ed Olney Story: Part I."

4   Forest Protection Division, Alberta Land and Forest Service, "Dozer Operations," in *Bull Dozer User Guide* (Hinton, AB: Hinton Training Centre, 2000).

5   Ibid.

6   Michael Duffy, "The Holt Tractor," *Firstworldwar.com*, last upated August 22, 2009, http://www.firstworldwar.com/atoz/holttractor.htm.

7   Ibid.

8   Ibid.

9   Memorandum to E.S. Huestis (director of forestry) from J. Harvie (deputy minister), May 2, 1951, file F-5, Alberta Forest Service.

10 P.J. Murphy, *History of Forest and Prairie Fire Control Policy in Alberta* (ENR report no. T/77, Edmonton: Alberta Energy and Natural Resources, 1985).

11 Canadian Institute of Forestry Standing Committee on Forest Fire, *The Forestry Chronicle*, special supplement, serial no. 104 (March 1951): 9–23.

12 Amiro et al., "Perspectives on Carbon Emissions," 388–90.

13 D. Sandink, "The Resilience of the City of Kelowna: Exploring Mitigation before, during and after the Okanagan Mountain Park Fire" (research paper series no. 45, Institute for Catastrophic Loss Reduction, Toronto, 2009).

14 J.L. Beverly and P. Bothwell, "Wildfire Evacuations in Canada 1980–2007," *Natural Hazards* 59 (2011): 571–96.

15 J. Podur and M. Wotton, "Will Climate Change Overwhelm Fire Management Capacity?" *Ecological Modeling* 221 (2010): 1301–09.

16 United Nations, "Fire Management—Global Assessment 2006: A Thematic Study Prepared in the Framework of the Global Forest Resources Assessment 2005" (FAO forestry paper 151, United Nations, Rome, IT, 2007), ftp://ftp.fao.org/docrep/fao/009/A0969E/A0969E00.pdf.

17 Dave Martell, "Forest Fire Management in Canada—We Have to Stop Behaving Like This," CIFFC National Fire Management Conversation (teleconference presentation, Toronto), January 24, 2007.

18 Ibid.

19 M.E. Alexander, C. Tymstra, and K.W. Frederick, "Incorporating Breaching and Spotting Considerations into PROMETHEUS—The Canadian Wildland Fire Growth Model" (Chisholm/Dogrib Fire Research Initiative Quicknote 6, Foothills Model Forest, Hinton, AB, November 2004), https://www.for.gov.bc.ca/ftp/!Project/FireBehaviour/CDFR_Qn6.pdf.

20 Ibid.

21 Ibid.

22 Department of the Interior, "Forestry and Irrigation," *Report of the Superintendent of Forestry and Irrigation*, part VII (sessional paper no. 25, Department of the Interior,

Forestry Branch, Ottawa, 1910), 24–34, http://cfs.nrcan.gc.ca/pubwarehouse/
pdfs/30797.pdf.

23   Ibid.

24   Ibid.

25   Backfire 2000 v United States, D. Mont., 9:03-cv-00198-DVM, doc. 85.

26   E.S. Huestis, *Report of the Director of Forestry. Annual Report of the Department of Lands and
Forests of the Province of Alberta for the Fiscal Year Ended March 31st 1957* (Edmonton: Queen's
Printer, 1958), 28–47.

27   P.J. Murphy et al., *The Alberta Forest Service, 1930– 2005: Protection and Management of Alberta's
Forests* (Edmonton: Department of Sustainable Resource Development, 2006).

28   Ibid.

29   Ibid.

30   Ibid.

31   Ibid.

32   "Canadair CL-215," accessed September 25, 2011, http://aircraft-list.com/db/
Canadair_CL-215/155/.

33   Ibid.

34   Bombardier, "Bombardier 415," accessed September 25, 2011,
http://www.bombardier.com/en/aerospace/amphibious-aircraft.html.

35   Martin Volk, "Martin Mars," July 25, 2008, http://www.seaplanes.org/mambo/
index.php?option=com_content&task=view&id=196&Itemid=245.

36   Ibid.

37   Ibid.

38   Glenn L. Martin Maryland Aviation Museum, "Martin Models 170, 193, and 199,"
accessed October 28, 2007, http://www.marylandaviationmuseum.org/history/
martin_aircraft/14_mars.html.

39   Volk, "Martin Mars."

40   Glenn L. Martin Maryland Aviation Museum, "Martin Models 170, 193, and 199."

41   Volk, "Martin Mars."

42   Ibid.

43   *Wikipedia*, s.v. "Martin JRM Mars," accessed September 24, 2014, http://en.wikipedia.
org/wiki/Martin_JRM_Mars.

44   Ibid.

45   Ibid.

46   Glenn L. Martin Maryland Aviation Museum, "The Glenn L. Maryland Aviation
Museum Loses in Bid to Acquire Historic Martin Mars Seaplane," accessed October
28, 2007, http://www.marylandaviationmuseum.org/mars/index.html.

47   Volk, "Martin Mars."

48   CBC News, "BC Water Bomber Arrives to Fight California Wildfires," October 25,
2007, http://www.cbc.ca/canada/british-columbia/story/2007/10/25/bc-
waterbomber.html?ref=rss.

49   Global Emergency Response, *Global Emergency Response: The Ilyushin Solution*, accessed
September 25, 2011, http://www.fire.uni-freiburg.de/emergency/ger.htm.

50   D. Graham-Rowe, "Airships Put a Damper on Forest Fires," *New Scientist*, June 22,
2002, http://scienceblog.com/community/older/2002/D/20024046.html.

51   Georg Breuer, *Weather Modification: Prospects and Problems* (Cambridge: Cambridge
University Press, 1980).

52 Alexander Linkewich, *Air Attack on Forest Fires: History and Techniques* (Red Deer, AB: D.W. Friesen, 1972).

53 Ibid.

54 C.C. Wilson and J.B. Davis, "Forest Fire Laboratory at Riverside and Fire Research in California: Past, Present, and Future" (general technical report PSW-105, Berkeley, CA: USDA Forest Service, Pacific Southwest Forest and Range Experiment Station, 1988).

55 G.P. Delisle and R.J. Hall, *Forest Fire History Maps of Alberta, 1931 to 1983* (Edmonton: Northern Forestry Centre, Canadian Forestry Service, 1987).

56 Ibid.

57 Alberta Environment and Sustainable Resource Development, "Wildfire Prevention and Enforcement," accessed September 24, 2014, http://wildfire.alberta.ca/ wildfire-prevention-enforcement/default.aspx.

58 S.K. Todd and H.A. Jewkes, "Wildland Fire in Alaska: A History of Organized Fire Suppression and Management in the Last Frontier" (University of Alaska Fairbanks, Agricultural and Forestry Experiment Station, bulletin no. 114, 2006).

59 FireSmart Canada, "What is FireSmart?" accessed September 24, 2014, https:// www.firesmartcanada.ca/what-is-firesmart; National Fire Protection Association, Firewise Communities, "About Firewise," accessed September 24, 2014, http://www. firewise.org/about.aspx.

60 Huestis, "Early Days in Alberta Forestry," 33–34.

61 Esperanza Investigation Team, *Esperanza Fire, Accident Investigation, Factual Report* (Riverside County, CA: USDA Forest Service and California Department of Forestry and Fire Protection, 2006), http://www.fire.ca.gov/fire_protection/downloads/ esperanza_00_complete_final_draft_05_01_2007.pdf.

62 R. Cathcart, "Arsonist Sentenced to Death for Killing 5 Firefighters," *New York Times*, June 5, 2009, http://www.nytimes.com/2009/06/06/us/06sentence.html.

63 California Department of Forestry and Fire Protection, Incident Command Team #8, "Last Fact Sheet for the Esperanza Fire from the Incident Base," October 31, 2006, http://bof.fire.ca.gov/pub/cdf/images/incidentfile161_325.pdf.

64 Randsc (forum member), "CNN Breaking News – 3 CA Wildfires LODDS," *Firehouse* (forum), October 26, 2006, http://www.firehouse.com/forums/archive/index. php/t-85006.html.

65 J.A. Blackwell and A. Tuttle, *California Fire Siege 2003: The Story* (Mare Island, CA: California Department of Forestry and Fire Protection, and United States Forest Service, Pacific Southwest Region, 2004), http://www.fire.ca.gov/ downloads/2003FireStoryInternet.pdf.

66 Ibid.

67 Ibid.

68 Ibid.

69 United States Government Accountability Office, "Technology Assessment: Protecting Structures and Improving Communications during Wildland Fires" (report to congressional requesters, GAO-05-380, 2005), http://www.gao.gov/ assets/160/157598.html.

70 California Department of Forestry and Fire Protection, "100 Feet of Defensible Space is the Law," September 27, 2007, http://www.fire.ca.gov/education_100foot.php.

71   Interagency Team, *June 2008 Fire Siege* (report prepared for the California Department of Forestry and Fire Protection) (Riverside, CA: United States Forest Service, Office of Emergency Services, and National Parks Service, 2008), http://www.fire.ca.gov/fire_protection/downloads/siege/2008/2008FireSiege_full-book_r6.pdf.

72   Ibid.

73   Ibid.

74   Del Walters, Lester A. Snow, and Arnold Schwarzenegger, *2009 Wildfire Activity Statistics* (n.p.: California Department of Forestry and Fire Protection, 2009), http://www.fire.ca.gov/downloads/redbooks/2009/2009_Redbook_Complete_Final.pdf.

75   National Wildfire Coordinating Group, *Wildfire Prevention Strategies* (PMS 455/NFES 1572), March 1998, http://www.nwcg.gov/pms/docs/wfprevnttrat.pdf.

76   B.J. Stocks et al., "Large Forest Fires in Canada, 1959–1997," *Journal of Geophysical Research* 107, D1 (2002), doi:10.1029/2001JD000484.

77   Ibid.

78   Canadian Council of Forest Ministers, *Canadian Wildland Fire Strategy Declaration* (Saskatoon, SK: Canadian Council of Forest Ministers, 2005).

CONCLUSION

1   W.J. Bloomberg, "Fire and Spruce," *Forestry Chronicle* 26, no. 2 (1950): 157–61.

2   M.D. Flannigan et al., "Future Area Burned in Canada," *Climatic Canada* 72 (2005): 1–16.

3   C.D. Krezek-Hanes et al., "Trends in Large Fires in Canada, 1959–2007" (Canadian Biodiversity: Ecosystem Status and Trends 2010, Technical Thematic Report no. 6., Canadian Councils of Resource Ministers, Ottawa, 2011), http://www.fire.uni-freiburg.de/inventory/database/Krezek-Hanes-2011-Large-Fires-Canada-1959-2007.pdf.

4   M-A. Parisien et al., "Scale-Dependent Controls on the Area Burned in the Boreal Forest of Canada, 1980–2005," *Ecological Applications* 21, no. 3 (2011): 789–805.

5   Canadian Wildland Fire Information System, CWFIS Datamart, accessed September 20, 2014, http://cwfis.cfs.nrcan.gc.ca/datamart. Fire polygon and point datasets are available by request through the CWFIS website.

6   Delisle and Hall, *Forest Fire History Maps of Alberta, 1931 to 1983*.

7   Government of Manitoba, "Busy Forest Fire Season Closes in Manitoba," news release, October 27, 2006, http://news.gov.mb.ca/news/index.html?archive=&item=223.

8   Victor Kafka (Parks Canada, Quebec City) and Rob Krauss (Saskatchewan Environment and Resource Management, Prince Albert), personal communication with the author.

9   B.D. Amiro et al., "Perspectives on Carbon Emissions from Canadian Forest Fires," *Forestry Chronicle* 78, no. 3 (2002): 388–90.

10   Government of Canada, Climate, "Daily Data Report for May 2011," accessed October 15, 2011, http://climate.weatheroffice.gc.ca/climate-Data/dailydata_e.html?timeframe=2&Prov=ALTA&StationID=31528&dlyRange=2006-01-01/2011-10-15&Year=2011&Month=5&Day=01.

11   Flat Top Complex Wildfire Review Committee, *Flat Top Complex, Final Report from the Flat Top Complex Wildfire Review Committee, Submitted to the Minister of Environment and Sustainable Resource Development, May 2012* (Edmonton: Alberta Environment and Sustainable Development, 2012), http://wildfire.alberta.ca/wildfire-prevention-enforcement/wildfire-reviews/documents/FlatTopComplex-WildfireReviewCommittee-May18-2012.pdf.

12   Ibid.

13   J.H. Williams, "The Mega-Fire Phenomenon: Risk Implications for Land Managers and Policy-Makers," Spring Forest Industry Lecture Series, March 4, 2010, University of Alberta, Edmonton, http://www.ales.ualberta.ca/rr/SeminarsandLectures/ForestIndustryLecture/Williams.aspx.

14   *Canadian Underwriter*, "Insured Damage from Slave Lake Wildfires Hits $700 Million-Mark," July 5, 2011, http://www.canadianunderwriter.ca/news/insured-damage-from-slave-lake-wildfires-hits-700-million-mark/1000505635/.

15   G. Thomson, "Extreme Weather Events Costing Alberta Millions," *Edmonton Journal*, August 6, 2001, http://www.edmontonjournal.com/news/Extreme+weather+events+costing+Alberta+millions/5215752/story.html.

16   IPCC, "Summary for Policymakers," in *Climate Change 2007: The Physical Science Basis. Contribution of Working Group I to the Fourth Assessment Report of the Intergovernmental Panel on Climate Change*, eds. S. Solomon et al. (Cambridge: Cambridge University Press, 2007), http://www.ipcc.ch/publications_and_data/publications_ipcc_fourth_assessment_report_wg1_report_the_physical_science_basis.htm.

17   "Global Temperatures," *The Carbon Brief*, accessed August 13, 2011, http://www.carbonbrief.org/profiles/global-temperatures.

18   G.A. McKinley et al., "Convergence of Atmospheric and North Atlantic Carbon Dioxide Trends on Multidecadal Timescales," *Nature Geoscience* 4 (2011): 606–10, doi:10.1038/ngeo1193.

19   Y. Pan et al., "A Large and Persistent Carbon Sink in the World's Forests," *Science*, August 19, 2011, doi:10.1126/science.1201609.

20   T.C. Bond et al., "Bounding the Role of Black Carbon in the Climate System: A Scientific Assessment," *Journal of Geophysical Research: Atmospheres* 118, no. 11 (2013): 5380–552, doi:10.1002/jgrd.50171

21   Rhett A. Butler, "The Asian Forest Fires of 1997–1988," accessed October 8, 2014, http://rainforests.mongabay.com/08indo_fires.htm.

22   Rhett A. Butler, "2007 Amazon Fires among Worst Ever," *Mongabay.com*, October 22, 2007, http://news.mongabay.com/2007/1021-amazon.html.

23   Statistics Canada, "Gross Domestic Product by Industry, May 2011," *The Daily*, July 29, 2011, http://www.statcan.gc.ca/daily-quotidien/110729/dq110729-eng.pdf.

24   H.W. Beall, "Fire Research in Canada's Federal Forestry Service—The Formative Years," in *The Art and Science of Fire Management: Proceedings of First Interior West Fire Council Annual Meeting and Workshop, October 24–27, 1988* (information report NOR-X-309, Forestry Canada, Northwest Region, Northern Forestry Centre, Edmonton, 1990), 14–19.

25   Ibid.

26   Murphy, *History of Forest and Prairie Fire Control Policy*.

27   J.G. Wright, "Forest-Fire Hazard Research: As Developed and Conducted at the Petawawa Forest Experiment Station" (forest fire hazard paper no. 2, Division of Forest Protection, Forest Service, Department of the Interior, Ottawa, 1932).

28   H.W. Beall, J.C. Macleod, and A. Potvin, "Forest Fire Danger in Midwestern Canada" (adapted from *Forest Fire Resolution*, n12), Dominion Forest Service, Canadian Department of Mines and Resources, Ottawa, 1949.

29   Beall, "Fire Research in Canada's Federal Forestry Service."

30   Ibid.

31   N.D. Burrows, "Predicting Blow-up Fires in the Jarrah Forest" (technical paper no. 12, Forests Department, South Perth, Western Australia, 1984), 3.

32   J.G. Wright and H.W. Beall, "The Application of Meteorology to Forest Fire Protection" (technical communication no. 4, Imperial Forestry Bureau, Oxford, 1945), 24.

33   Canadian Wildland Fire Strategy Assistant Deputy Ministers Task Group, *Canadian Wildland Fire Strategy: A Vision for an Innovative and Integrated Approach to Managing the Risks: Report to the Canadian Council of Forest Ministers* (Edmonton: Canadian Council of Forest Ministers, 2005).

34   LaFoy, "The 1950 Chinchaga River Fire."

35   Alberta Department of Lands and Forests, *Naylor Hills Fire Report*.

BIBLIOGRAPHY

The 75th Anniversary of the Great Fire of 1922 Committee. *The Great Fire of 1922*. Haileybury, ON: Haileybury Heritage Museum, 1999, repr. May 2000.

Alberta Department of Lands and Forests. *Fire 28A-2 Report*. Peace River: Forest Service, 1958.

———. *Naylor Hills Fire Report*. Peace River: Forest Service, 1950.

———. *Outlaw Fire Report*. Peace River: Forest Service, 1950.

———. *Rat Fire Report*. Peace River: Forest Service, 1950.

———. *Willow Swamp Fire Report*. Peace River: Forest Service, 1950.

Alberta Department of Lands and Mines. *Boundry Fire Report*. Peace River: Forest Service, 1950.

Alberta Forest Products Association. *Alberta's Forests: A Renewable Story*. Edmonton: Alberta Forest Productions Association, n.d. Accessed September 24, 2014. https://www.albertaforestproducts.ca/sites/default/files/downloads/Renewable%20Story.pdf.

Alexander, Martin E. "'Lest We Forget': Canada's Major Wildland Fire Disasters of the Past, 1825–1938." In *Proceedings of 3rd Fire Behavior and Fuels Conference, October 25–29, 2010, Spokane, Washington, US*. Birmingham, AL: International Association of Wildland Fire, 2010. https://www.firesmartcanada.ca/images/uploads/resources/Alexander-Lest-We-Forget.pdf.

Alexander, M.E., C. Tymstra, and K.W. Frederick. "Incorporating Breaching and Spotting Considerations into PROMETHEUS—The Canadian Wildland Fire Growth Model." Chisholm/Dogrib Fire Research Initiative Quicknote 6, Foothills Model Forest, Hinton, AB, November 2004. https://www.for.gov.bc.ca/ftp/!Project/FireBehaviour/CDFR_Qn6.pdf.

Amiro, B.D., M.D. Flannigan, B.J. Stocks, and B.M. Wotton. "Perspectives on Carbon Emissions from Canadian Forest Fires." *Forestry Chronicle* 78, no. 3 (2002): 388–90.

Amiro, B.D., J.B. Todd, B.M. Wotton, K.A. Logan, M.D. Flannigan, B.J. Stocks, J.A. Mason et al. "Direct Carbon Emissions from Canadian Forest Fires, 1959–1999." *Canadian Journal of Forest Research* 31 (2001): 512–25.

Ammann, Harriet. *Wildfire Smoke: A Guide for Public Health Officials*. Sacramento, CA: Office of Environmental Health Hazard Assessment, 2001.

Annand, Amanda Dawn. "The 1910 Fires in Alberta's Foothill and Rocky Mountain Regions." Hon. BSc. thesis, University of Victoria, Victoria, BC, 2010.

Apps, M.J., W.A. Kurz, R.J. Luxmore, L.O. Nilsson, R.A. Sedjo, R. Schmidt, L.G. Simpson, and T.S. Vinson. "Boreal Forests and Tundra." *Water, Air, and Soil Pollution* 70 (1993): 39–53.

Badash, Lawrence. *A Nuclear Winter's Tale: Science and Politics in the 1980s*. Cambridge, MA: MIT Press, 2009.

Ballantyne, E.E. "Rabies Control in Alberta." *Canadian Journal of Comparative Medicine and Veterinary Science* 20, no. 1 (1956): 21–30.

Barlowe, Jerome, and William Roye et al. *Rede Me and Be Nott Wrothe, for I Saye No Thinge But Trothe*, edited by D.H. Parker. Toronto: University of Toronto Press, 1992. Originally published in 1528.

Beall, H.W. "Fire Research in Canada's Federal Forestry Service—The Formative Years." In *The Art and Science of Fire Management: Proceedings of First Interior West Fire Council Annual Meeting and Workshop, October 24–27, 1988*. Information Report NOR-X-309, Forestry Canada, Northwest Region, Northern Forestry Centre, Edmonton, 1990.

Beverly, J.L., and P. Bothwell. "Wildfire Evacuations in Canada 1980–2007." *Natural Hazards* 59 (2011): 571–96.

Birch Hills Historical Society. *Grooming the Grizzly: A History of Wanham and Area*, edited by W. Tansen. Winnipeg, MB: Inter-Collegiate Press, 1982.

Blackwell, J.A., and A. Tuttle, *California Fire Siege 2003: The Story*. Mare Island, CA: California Department of Forestry and Fire Protection, and United States Forest Service, Pacific Southwest Region, 2004. http://www.fire.ca.gov/downloads/2003FireStoryInternet.pdf.

Blefgen, T.F. *Report of the Director of Forestry. Annual Report of the Department of Lands and Mines of the Province of Alberta for the Fiscal Year Ended March 31st 1935*. Edmonton: King's Printer, 1936.

———. *Report of the Director of Forestry. Annual Report of the Department of Lands and Mines of the Province of Alberta for the Fiscal Year Ended March 31st 1937*. Edmonton: King's Printer, 1938.

———. *Report of the Director of Forestry. Annual Report of the Department of Lands and Mines of the Province of Alberta for the Fiscal Year Ended March 31st 1943*. Edmonton: King's Printer, 1944.

———. *Report of the Director of Forestry. Annual Report of the Department of Lands and Mines of the Province of Alberta for the Fiscal Year Ended March 31st 1945*. Edmonton: King's Printer, 1946.

———. *Report of the Director of Forestry. Annual Report of the Department of Lands and Mines of the Province of Alberta for the Fiscal Year Ended March 31st 1950*. Edmonton: King's Printer, 1951.

———. *Report of the Director of Forestry. Annual Report of the Department of Lands and Mines of the Province of Alberta for the Fiscal Year Ended March 31st 1955*. Edmonton: King's Printer, 1956.

Bloomberg, W.J. "Fire and Spruce." *Forestry Chronicle* 26, no. 2 (1950): 157–61.

Bohren, Craig. F. *Clouds in a Glass of Beer: Simple Experiments in Atmospheric Physics*. Mineola, NY: Dover Publications, 1987.

Bohren, Craig F., and Donald R. Huffman. *Absorption and Scattering of Light by Small Particles*. New York: John Wiley & Sons, 1983.

Bond, T.C., S.J. Doherty, D.W. Fahey, P.M. Forster, T. Berntsen, B.J. DeAngelo, M.G. Flanner et al. 2013. "Bounding the Role of Black Carbon in the Climate System: A Scientific

Assessment." *Journal of Geophysical Research: Atmospheres* 118, no. 11 (2013): 5380–552. doi:10.1002/jgrd.50171.

Breuer, Georg. *Weather Modification: Prospects and Problems*. Cambridge: Cambridge University Press, 1980.

British Columbia Department of Lands and Forests. *Want Lake Fire Report*. Prince George: Forest Service, 1950.

———. *Whisp Fire Report*. Fort St. John: Forest Service, 1950.

Bull, G.A. "Blue Sun and Moon." *Marine Observer* 21, no. 153 (1951): 167–69.

Burrows, N.D. "Predicting Blow-up Fires in the Jarrah Forest." Technical Paper no. 12. Forests Department, South Perth, Western Australia, 1984.

California Department of Forestry and Fire Protection, Incident Command Team #8. "Last Fact Sheet for the Esperanza Fire from the Incident Base." Fact Sheet. October 31, 2006. http://bof.fire.ca.gov/pub/cdf/images/incidentfile161_325.pdf.

Canada, Department of the Interior. "Annual Report of the Department of the Interior for the Year 1897." Sessional Papers, no. 13. A., Department of the Interior, Ottawa, 1898.

Canadian Council of Forest Ministers. *Canadian Wildland Fire Strategy Declaration*. Saskatoon, SK: Canadian Council of Forest Ministers, 2005.

Canadian Institute of Forestry. *Forest Fire Protection in Alberta: A Review and Recommendations*. N.p.: Canadian Institute of Forestry, Rocky Mountain Section, 1954.

Canadian Institute of Forestry Standing Committee on Forest Fire. *The Forestry Chronicle*. Special Supplement, serial no. 104 (March 1951): 9–23.

Canadian Interagency Forest Fire Centre. 2011. Current Situation Reports. September 22, 2010, 1500h. [Last report for the 2010 fire season]. Accessed March 27, 2011. http://www.ciffc.ca/firewire/current.php.

Canadian Medical Association. *No Breathing Room: National Illness Costs of Air Pollution. Summary Report*. Ottawa: Canadian Medical Association, August 2008.

Canadian Pulp and Paper Association. *Forest Fire Fighters Guide*. Montreal: Canadian Pulp and Paper Association, Woodlands Section, 1949.

Canadian Wildland Fire Strategy Assistant Deputy Ministers Task Group. *Canadian Wildland Fire Strategy: A Vision for an Innovative and Integrated Approach to Managing the Risks. Report to the Canadian Council of Forest Ministers*. Edmonton: Canadian Council of Forest Ministers, 2005.

CBC-Radio. *Round Up*, featuring Gerrard Faye. Originally broadcast September 28, 1950. CBC Digital Archives. Accessed March 28, 2011. http://www.cbc.ca/archives/categories/environment/natural-disasters/blue-sun-shines-over-england.html.

Chisholm Fire Review Committee. *Chisholm Fire Review Committee Final Report, Submitted to the Minister of Alberta Sustainable Resource Development, October 2001*. Edmonton: Alberta Environment and Sustainable Resource Development, 2001.

Crutzen, P.J., and J.W. Birks. "The Atmosphere after Nuclear War: Twilight at Noon." *The Aftermath: The Human and Ecological Consequences of Nuclear War*, edited by J. Peterson. Special issue of *Ambio* 11, no. 2–3 (1982): 76–83.

DeBano, L.F. "Chaparral Soils." In *Symposium on Living with the Chaparral: Proceedings*, edited by Murray Rosenthal, 19–26. San Francisco: Sierra Club, 1974.

DeBano, Leonard F., Daniel G. Neary, and Peter F. Ffolliott. *Fire's Effects on Ecosystems*. New York: John Wiley & Sons, 1998.

de Groot, W.J., R. Landry, W.A. Kurz, K.R. Anderson, P. Englefield, R.H. Fraser, R.J. Hall et al. "Estimating Direct Carbon Emissions from Canadian Wildland Fires." *International Journal of Wildland Fire* 16 (2007): 593–606.

Delisle, G.P., and R.J. Hall. *Forest Fire History Maps of Alberta, 1931 to 1983*. Edmonton: Northern Forestry Centre, Canadian Forestry Service, 1987.

Department of the Interior. "Forestry and Irrigation." In *Report of the Superintendent of Forestry and Irrigation*. Part VII. Sessional Paper no. 25, Department of the Interior, Forestry Branch, Ottawa, 1910. http://cfs.nrcan.gc.ca/pubwarehouse/pdfs/30797.pdf.

Department of Transport, Meteorological Division. *Monthly Record of Meteorological Observations in Canada* (August 1950).

———. *Monthly Record of Meteorological Observations in Canada* (September 1950).

Des Voeux, Henry Antoine. "Fog and Smoke." Paper presented at the Royal Institute of Public Health, London Congress Meeting, July 1905, London, UK.

Dion, Joseph F. *My Tribe, the Crees*. Calgary, AB: Glenbow Museum, Calgary, 1979.

Doucet, J.A. *Timber Conditions in the Smoky River Valley and Grande-Prairie Country*. Forestry Branch Bulletin no. 53, Department of the Interior, Ottawa, 1915.

Doyle, J.A. "A Survey of Forest Fire Causes and Suggested Corrective Methods." *Forestry Chronicle* 27, no. 4 (1951): 335–48.

Eastern Rockies Forest Conservation Board. *Annual Report of the Eastern Rockies Forest Conservation Board for the Fiscal Year 1949–50*. Calgary: Eastern Rockies Forest Conservation Board, 1950.

———. *Annual Report of the Eastern Rockies Forest Conservation Board for the Fiscal Year 1950–51*. Calgary: Eastern Rockies Forest Conservation Board, 1951.

Egan, Timothy. *The Big Burn: Teddy Roosevelt and the Fire That Saved America*. Boston: Houghton Mifflin Harcourt, 2009.

Elsley, E.M. "Alberta Forest-Fire Smoke—24 September 1950." *Weather* 6, no. 1 (1951): 22–26.

Environment Canada. *Canadian Normals: Precipitation 1941–1970*. Vol. 2. Downsview, ON: Atmospheric Environment, 1973.

Esperanza Investigation Team. *Esperanza Fire, Accident Investigation, Factual Report*. Riverside County, CA: USDA Forest Service and California Department of Forestry and Fire Protection, 2006. http://www.fire.ca.gov/fire_protection/downloads/esperanza_00_complete_final_draft_05_01_2007.pdf.

Evans, W.G. "Perception of Infrared Radiation from Forest Fires by Melanophila Acuminata de Geer (Buprestidae, Coleoptera)." *Ecology* 47 (1966): 1061–65.

Filmon, Gary, and Firestorm 2003 Provincial Review Team. *Firestorm 2003 Provincial Review*. Vancouver: Government of British Columbia, 2004. http://bcwildfire.ca/History/ReportsandReviews/2003/FirestormReport.pdf.

Flannigan, M.D., K.A. Logan, B.D. Amiro, W.R. Skinner, and B.J. Stocks. "Future Area Burned in Canada." *Climatic Canada* 72 (2005): 1–16.

Flat Top Complex Wildfire Review Committee. *Flat Top Complex, Final Report from the Flat Top Complex Wildfire Review Committee, Submitted to the Minister of Environment and Sustainable Resource Development, May 2012*. Edmonton: Alberta Environment and Sustainable Development, 2012. http://wildfire.alberta.ca/wildfire-prevention-enforcement/wildfire-reviews/documents/FlatTopComplex-WildfireReviewCommittee-May18-2012.pdf

Forest Protection Division, Alberta Land and Forest Service. "Dozer Operations." In *Bull Dozer User Guide*. Hinton, AB: Hinton Training Centre, 2000.

Forestry Canada Fire Danger Group. *Development and Structure of the Canadian Forest Fire Behavior Prediction System*. Information Report ST-X-3. Ottawa: Forestry Canada, 1992.

Fromm, M.D., and R. Servranckx. "Transport of Forest Fire Smoke above the Troposphere by Supercell Convection." *Geophysical Research Letters* 30 (2002): 1542–45.

Fromm, M.D., J. Alfred, K. Hoppel, J. Hornstein, R. Bevilacqua, E. Shettle, R. Servranckx, Z. Li, and B. Stocks. "Observations of Boreal Forest Fire Smoke in the Stratosphere by POAM III, SAGE II and Lidar in 1998." *Geophysical Research Letters* 27 (2000): 1407–10.

Government of Alberta. *Alberta Forest Inventory*. Edmonton: Department of Lands and Forests, Forest Service, 1968.

———. *The Alberta Story: An Authentic Report on Alberta's Progress, 1935–1952*. Edmonton: Publicity Bureau, Department of Economic Affairs, 1952.

———. *Fallen Firefighter Memorial Dedication Ceremony*. September 20, 2008. Hinton, AB: Hinton Training Centre. DVD.

———. *Forest Cover Map of Alberta*. Edmonton: Department of Lands and Forests, Forest Service, 1957.

———. *Map of Areas (1,000 Acres and over) Devastated by Forest Fires 1938 to 1949 Inclusive*. Edmonton: Department of Lands and Forests, 1949.

Government of British Columbia. *Report of Forest Service*. Victoria: Department of Lands and Forests, 1951.

Government of Canada. *Forest Fire Losses in Canada 1949*. Ottawa: Forestry Branch, Forest Research Division, Department of Resources and Development, 1950.

———. *Report of the Department of Resources and Development for the Fiscal Year Ended March 31, 1950*. Ottawa: Forestry Branch, 1950.

———. *Report of the Director of Forestry For The Fiscal Ended March 31, 1920*. Ottawa: Department of the Interior, 1921. http://cfs.nrcan.gc.ca/bookstore_pdfs/30885.pdf.

———. *Review of Fire Season 1919 on Dominion Forest Reserves & Fire Ranging Districts*. Ottawa: Dominion Forestry Branch, 1919.

Government of New Brunswick. *Great Miramichi Fire*. Fredericton: New Brunswick Natural Resources, 2009. http://web.archive.org/web/20101013094259/http://www.gnb.ca/0079/miramichi_fire-e.asp.

Grant, Jack. Interview by Peter J. Murphy (January 26, 1999). In *Alberta Forest Service History Project: 2005 – 75th Anniversary*. Edmonton: Alberta Land and Forest Service, 1999.

Gratkowski, H.J. "Heat As a Factor in Germination of Seeds of Ceanothus velutinus var. laevigatus T. & G." PhD diss., Oregon State University, Corvallis, 1962.

Hanson, W.R. *History of the Eastern Rockies Forest Conservation Board, 1947–1973*. Calgary, AB: Eastern Rockies Forest Conservation Board, 1973.

Hardin, Mary, and Ralph Kahn. "Aerosols and Climate Change." *Earth Observatory*. Last updated November 1, 2010. http://earthobservatory.nasa.gov/Features/Aerosols/what_are_aerosols_1999.pdf.

Harper, K.C. *Weather and Climate: Decade by Decade*. New York: Facts on File, 2007.

Hart, S. "Beetlemania: An Attraction to Fire." BioScience 48, no. 1 (1998): 3–5.

Henderson, D.E. "Prescribed Burns and Wildfires in Colorado: Impacts of Mitigation Measures on Indoor Air Particulate Matter." *Journal of the Air and Waste Management Association* 55, no. 10 (2005): 1516–26.

Henderson, Sarah, and Michael Brauer. "Fire Smoke and Human Health." *Canadian Smoke Newsletter*, Spring 2008, 6–7. http://www.sk.lung.ca/canadian_smoke_newsletters/spring_2008.pdf.

Hind, Henry Youle. *Explorations in the Interior of the Labrador Peninsula: The Country of the Montagnais and Nasquapee Indians*. Vol. 1. London: Longman, Green, Longman, Roberts, and Green, 1863.

Holdsworth, G. "Evidence for a Link between Atmospheric Thermonuclear Detonations and Nitric Acid. *Nature* 324 (1986): 551–53.

Hope, Sharon, and ASIM Ltd. "Community Wildland Fire Protection Plan for the District of Tumbler Ridge." Report submitted to Dan Golob, Fire Chief, Tumbler Ridge Fire Department, Tumbler, BC, August 2006. http://www.tumblerridge.ca/Portals/0/ Community%20Wildland%20Fire%20Protection%20Plan_Support%20Document%20 09-04_RFP.pdf.

Huestis, E.S. "Early Days in Alberta Forestry." Talk given to new Alberta Forest Service forest officers during an orientation course at the Forest Technology School in Hinton, AB, January 31, 1972. Transcribed and edited by P.J. Murphy, July 1986.

————. *Report of the Assistant Director of Forestry. Annual Report of the Department of Lands and Mines of the Province of Alberta for the Fiscal Year Ended March 31st 1948*. Edmonton: King's Printer, 1949.

————. *Report of the Director of Forestry. Annual Report of the Department of Lands and Mines of the Province of Alberta for the Fiscal Year Ended March 31st 1949*. Edmonton: King's Printer, 1950.

————. *Report of the Director of Forestry. Annual Report of the Department of Lands and Forests of the Province of Alberta for the Fiscal Year Ended March 31st 1950*. Edmonton: King's Printer, 1951.

————. *Report of the Director of Forestry. Annual Report of the Department of Lands and Mines of the Province of Alberta for the Fiscal Year Ended March 31st 1951*. Edmonton: King's Printer, 1952.

————. *Report of the Director of Forestry. Annual Report of the Department of Lands and Mines of the Province of Alberta for the Fiscal Year Ended March 31st 1952*. Edmonton: King's Printer, 1953.

————. *Report of the Director of Forestry. Annual Report of the Department of Lands and Forests of the Province of Alberta for the Fiscal Year Ended March 31st 1957*. Edmonton: Queen's Printer, 1958.

————. *Report of the Director of Forestry. Annual Report of the Department of Lands and Forests of the Province of Alberta for the Fiscal Year Ended March 31st 1961*. Edmonton: Queen's Printer, 1962.

Interagency Team. *June 2008 Fire Siege*. (Report prepared for the California Department of Forestry and Fire Protection). Riverside, CA: United States Forest Service, Office of Emergency Services, and National Parks Service, 2008. http://www.fire.ca.gov/fire_protection/ downloads/siege/2008/2008FireSiege_full-book_r6.pdf.

Intergovernmental Panel on Climate Change. "Summary for Policymakers." In *Climate Change 2007: The Physical Science Basis. Contribution of Working Group I to the Fourth Assessment Report of the Intergovernmental Panel on Climate Change*, edited by S. Solomon, D. Qin, M. Manning, Z. Chen, M. Marquis, K.B. Averyt, M. Tignor, and H.L. Miller. Cambridge: Cambridge University Press, 2007. http://www.ipcc.ch/publications_and_data/publications_ipcc_fourth_ assessment_report_wg1_report_the_physical_science_basis.htm.

Jackson, Frank. *A Candle in the Grub Box: A Struggle for Survival in the Northern Wilderness. The Story of Frank Jackson as Told to and Transcribed by Sheila Douglass*. Victoria, BC: Shires Books, 1977.

Jackson, Mary Percy. *Suitable for the Wilds: Letters from Northern Alberta 1929–1931*. Calgary: University of Calgary Press, 2006.

Jackson, Mary Percy, and Cornelia Lehn. *The Homemade Brass Plate: The Story of a Pioneer Doctor in Northern Alberta as Told to Cornelia Lehn*, edited by Janice Dickin. Sardis, BC: Cedar-Cott Enterprise, 1988.

Janz, B., and N. Nimchuk. "The 500 mb Anomaly Chart—A Useful Fire Management Tool." In *Proceedings of the Eighth Conference on Fire and Forest Meteorology, April 29–May 2, Detroit, MI*, 233–338. Bethesda, MD: Society of American Foresters, 1985.

Janzen, S. "The Burning North: A History of Fire and Fire Protection in the NWT." MA thesis, University of Alberta, Edmonton, 1990.

Johnson, L.E. "Commemorating the 80th Anniversary of the Fires of 1918." *Minnesota History Interpreter* 26, no. 6 (1998): 1–2.

Johnston, Fay H., Sarah B. Henderson, Yang Chen, James T. Randerson, Miriam Marlier, Ruth
S. DeFries, Patrick Kinney et al."Estimated Global Mortality Attributable to Smoke from
Landscape Fires." *Environmental Health Perspectives* 120, no. 5 (2012): 659–701. http://ehp.
niehs.nih.gov/1104422/

Joint Action Implementation Coordinating Committee. *Report to the Council of Ministers of Environment:
An Update in Support of the Canada-Wide Standards for Particulate Matter and Ozone*. N.p.: Joint Action
Implementation Coordinating Committee, 2005. http://www.ccme.ca/files/Resources/
air/pm_ozone/pm_03_update_2005_e.pdf.

Joy, George C. "The Conflagration: Western Possibilities of Sweeping Fires Like Those of
Minnesota and Canada." Address delivered at Annual Conference, Western Forestry and
Conservation Association, Portland, November 25, 1922. Portland: Oregon State Board of
Forestry, 1922.

Kalnay, E., M. Kanamitsu, R. Kistler, W. Collins, D. Deaven, L. Gandin, M. Iredell, S. Saha et
al. "The NCEP/NCAR 40-Year Reanalysis Project." *Bulletin of the American Meteorological Society* 77
(1996): 437–71.

Keg River History Book Committee. *"Way Out Here": History of Carcajou, Chinchaga, Keg River, Paddle Prairie,
Twin Lakes*. Altona, MB: Friesen Printers, 1994.

Krezek-Hanes, C.C., F. Ahern, A. Cantin, and M.D. Flannigan. "Trends in Large Fires in Canada,
1959–2007." *Canadian Biodiversity: Ecosystem Status and Trends 2010*. Technical Thematic Report no.
6., Canadian Councils of Resource Ministers, Ottawa, 2011. http://www.fire.uni-freiburg.
de/inventory/database/Krezek-Hanes-2011-Large-Fires-Canada-1959-2007.pdf.

Kurz, W.A., and M.J. Apps. "Developing Canada's National Forest Carbon Fluxes in the Canadian
Forest Sector." *Ecological Applications* 9 (2006): 526–47. doi:10.1007/S11027-006-1006-6.

Lafleur, Laurence. "Sauce for the Gander." Question and Answer Column. *Sky & Telescope*, July 1943.

LaFoy, Frank. "The 1950 Chinchaga River Fire. Excerpts from an Interview with Frank L.
LaFoy, former Forest Ranger," by John Frank, undergraduate forestry student, Winter
1977. Directed study for Peter J. Murphy. Edited by P.J. Murphy and Cordy Tymstra
and transcribed by Judy Jacobs. Faculty of Agriculture and Forestry, University of
Alberta, Edmonton.

Lavoué, David. "Quebec Wildfires Spread Smoke to Eastern Canada and New England." *Canadian
Smoke Newsletter*, Fall/Winter 2010, 15–17. http://www.sk.lung.ca/canadian_smoke_
newsletters/fall_winter_2010.pdf.

Lavoué, D., and B.J. Stocks. "Emissions of Air Pollutants by Canadian Wildfires from 2000 to
2004." *International Journal of Wildland Fire* 20, no. 1 (2011): 17–34. doi:10.1071/WF08114.

Lavoué, D., S.L. Gong, and B.J. Stocks. "Modelling Emissions from Canadian Wildfires: A
Case Study of the 2002 Quebec Forest Fires." *International Journal of Wildland Fire* 16, no. 6
(2007): 649–63.

Leonard, David W. *The Last Great West: The Agricultural Settlement of the Peace River County to 1914*. Calgary,
AB: Detselig Enterprises Ltd., 2005.

Levy, H., W.J. Moxim, A.A. Klonecki, and P.S. Kasibhatla. "Simulated Tropospheric NOx: Its
Evaluation, Global Distribution and Individual Source Contributions." *Journal of Geophysical
Research–Atmospheres* 104 (1999): 26279–306. doi:10.1029/1999JD900442.

Lewis, H.T. "A Time for Burning." Occasional publication no. 17, Boreal Institute for Northern
Studies, University of Alberta, Edmonton, 1982.

Linkewich, Alexander. *Air Attack on Forest Fires: History and Techniques*. Red Deer, AB: D.W.
Friesen, 1972.

Lyman, H. "Smoke from Minnesota Forest Fires." *Monthly Weather Review* 46, no. 11 (1919): 506–09.

MacGibbon, D.A. *The Canadian Grain Trade 1931–1951*. Toronto: University of Toronto Press, 1952.

Maini, J.S. "Silvics and Ecology of Populus in Canada." In *Growth and Utilization of Poplars in Canada*, edited by J.S. Maini and J.H. Cayford, 20–69. Ottawa: Canada Department of Forestry and Rural Development, 1968.

Marley, N.A., J.S. Gaffney, P.J. Drayton, M.M. Cunningham, C. Mielcarek, R. Ravelo, and C. Wagner. "Residence Times of Fine Tropospheric Aerosols as Determined by [210]Pb Progeny." In *Symposium on Atmospheric Chemistry Issues in the 21st Century*, 1–5. Boston: American Meteorological Society, 2000.

Martell, Dave. "Forest Fire Management in Canada—We Have to Stop Behaving Like This." CIFFC National Fire Management Conversation. Teleconference presentation, Toronto, January 24, 2007.

Martynowych, Orest T. "The Ukrainian Bloc Settlement in East Central Alberta, 1890–1930: A History." Occasional Paper no. 10, Historic Sites Service, Alberta Culture, Edmonton, 1985.

Matheson, Shirlee Smith. *Lost: True Stories of Canadian Aviation Tragedies*. Calgary, AB: Fifth House Ltd., 2005.

McClement, Fred. *The Flaming Forests*. Toronto: McClelland and Stewart, 1969.

McClure, R. "Washington's 'Awful Conflagration'—The Yacolt Fire of 1902." *Fire Management Today* 65, no. 1 (2005): 24–27.

McDonough, W.T. "Quaking Aspen-Seed Germination and Early Seedling Growth." Research Paper, INT-234, Intermountain Forest and Range Experiment Station, Ogden, UT, 1979.

McKee, J. "Fire Protection in British Columbia." In *Proceedings of the 55th Session of the Pacific Logging Congress. Loggers Handbook*. Vol. 25, 94–96. Portland, OR: Pacific Logging Congress, 1965.

McKinley, G.A., A.R. Fay, T. Takahashi, and N. Metzl. "Convergence of Atmospheric and North Atlantic Carbon Dioxide Trends on Multidecadal Timescales." *Nature Geoscience* 4 (2011): 606–10. doi:10.1038/ngeo1193.

McMurry, Erin R., Stambaugh, Michael C., Guyette, Richard P., and Dey, Daniel C. "Fire Scars Reveal Source of New England's 1780 Dark Day." *International Journal of Wildland Fire* 16 (2007): 266–70.

McRae, D. "Prescribed Burning for Stand Conversion in Budworm-Killed Balsam Fir: An Ontario Case History." *Forestry Chronicle* 62, no. 2 (1986): 96–100.

Meinel, Aden. *Sunsets, Twilights, and Evening Skies*. New York: Cambridge University Press, 1983.

Morris, William G. "Forest Fires in Western Oregon and Western Washington." *Oregon Historical Quarterly* 35 (1934): 313–39.

Murphy, Peter J. "Fire Research at the University of Alberta." In *Proceedings of Fire Ecology in Resource Management Workshop, Dec. 6–7, 1977, Edmonton*, compiled by D.E. Dubé. Information Report NOR-X-210, Canadian Forest Service, Northern Forest Research Centre, Edmonton, 1978.

———. "'Following the Base of the Foothills': Tracing the Boundaries of Jasper Park and Its Adjacent Rocky Mountain Forest Reserve." In *Culturing Wilderness in Jasper National Park: Studies in Two Centuries of Human History in the Upper Athabasca River Watershed*, edited by I.S. MacLaren, 71–121. Edmonton: University of Alberta Press, 2007.

———. *History of Forest and Prairie Fire Control Policy in Alberta*. ENR Report no. T/77. Edmonton: Alberta Energy and Natural Resources, 1985.

Murphy, Peter J., and Cordy Tymstra. "The 1950 Chinchaga River Fire in the Peace River Region of British Columbia/Alberta: Preliminary Results of Simulating Forward Spread Distances." In *Proceedings of the Third Western Region Fire Weather Committee Scientific and Technical Seminar, February 4, 1986*, edited by Martin E. Alexander, 20–30. Study NOR-5-05, (NOR-5-19), file report no. 15. Canadian Forestry Service, Edmonton, 1986. http://cfs.nrcan.gc.ca/pubwarehouse/pdfs/24290.pdf.

Murphy, P.J., R.E. Stevenson, D. Quintilio, and S. Ferdinand. *The Alberta Forest Service, 1930–2005: Protection and Management of Alberta's Forests*. Edmonton: Alberta Sustainable Resource Development, Government of Alberta, 2006.

Mutch, R.W. "Wildland Fires and Ecosystems—A Hypothesis." *Ecology* 51 (1970): 1046–51.

Natural Resources Canada. *The Atlas of Canada, Boreal Forest*. Accessed March 26, 2011. http://atlas. nrcan.gc.ca/site/english/learningresources/theme_modules/borealforest/index.html.

———. *The State of Canada's Forests. Annual Report 2010*. Ottawa: Natural Resources Canada, Canadian Forest Service, 2010.

National Wildfire Coordinating Group. *Wildfire Prevention Strategies*. PMS 455/NFES 1572, March 1998. http://www.nwcg.gov/pms/docs/wfprevnttrat.pdf.

Ontario Medical Association. *The Illness Costs of Air Pollution: 2005–2026 Health and Economic Damage Estimates*. Toronto: Ontario Medical Association, June 2005. https://www.oma.org/ Resources/Documents/e2005HealthAndEconomicDamageEstimates.pdf.

Ontario Ministry of the Environment. *Air Quality in Ontario 2002 Report*. Toronto: Ontario Ministry of the Environment, Environmental Monitoring and Reporting Branch, 2002. http://www. airqualityontario.com/downloads/AirQualityInOntarioReport2002.pdf.

Ottmar, Roger D. "Smoke Source Characteristics." In *Smoke Management Guide for Prescribed and Wildland Fire, 2001 Edition*, edited by Colin C. Hardy, Roger D. Ottmar, Janice L. Peterson, John E. Core, and Paula Seamon, 89–106. Rep. PMS 420-2/NFES 1279, National Wildfire Coordination Group, Boise, ID, 2001.

Page, S.E., F. Siegert, J.O. Rieley, H-D.V Boehm, and A. Jaya. "Carbon Released during Peatland Fires in Central Lalimantan, Indonesia in 1997." *Nature* 420 (2002): 61–65.

Pan, Y., R.A. Birdsey, J. Fang, R. Houghton, P.E. Kauppi, W.A. Kurz, O.L. Phillips et al. "A Large and Persistent Carbon Sink in the World's Forests." *Science*, August 19, 2011. doi:10.1126/ science.1201609.

Parisien, M-A, S.A. Parks, M.A. Krawchuck, M.D. Flannigan, L.M. Bowman, and M.A. Moritz. "Scale-Dependent Controls on the Area Burned in the Boreal Forest of Canada, 1980–2005." *Ecological Applications* 21, no. 3 (2011): 789–805.

Parliament of Victoria. *2009 Victorian Bushfires Royal Commission. Final Report Summary*. Victoria, AU: Government Printer for the State of Victoria, 2009.

Parminter, John. *1950 Fire Report Summaries*. Victoria: British Columbia Forest Service, 1981.

Partners in Protection. *FireSmart: Protecting Your Community from Wildfire*. 2nd ed. Edmonton: Partners in Protection, 2003.

Payne, N.J., B.J. Stocks, A. Robinson, M. Wasey, and J.W. Strapp. "Combustion Aerosol from Experimental Crown Fires in a Boreal Forest Jack Pine Stand." *Canadian Journal of Forest Research* 34 (2004): 1627–33.

Penndorf, Rudolf. "On the Phenomenon of the Colored Sun, Especially the 'Blue' Sun of September 1950." Geophysical Research Papers no. 20, Geophysics Research Directorate, Air Force Cambridge Research Center, Cambridge, MA, 1953.

Pesic, P. "A Simple Explanation of Blue Suns and Moons." *European Journal of Physics* 29 (2008): N31–N36.

Philip, Tom. "Pioneering Peace Medicine: How Dr. Mary Percy Jackson Conquered the Wild and Healed the Sick." *Alberta Report*, December 26, 1983, 30–33.

Platt, C.F. "Fire Suppression." In *Complete Set of Papers for the Alberta First Natural Resources Conference, Theme: Inventory*. Edmonton: Government of Alberta, 1959.

Plummer, F.G. "Forest Fires: Their Causes, Extent and Effects, with a Summary of Recorded Destruction and Loss." USDA Forest Service Bulletin no. 117, Washington, DC, 1912.

Podur, J., and M. Wotton. "Will Climate Change Overwhelm Fire Management Capacity?" *Ecological Modeling* 221 (2010): 1301–09.

Porch, W.M. "Blue Moons and Large Fires." *Applied Optics* 28, no. 10 (1989): 1778–84.

Porch, W.M., C.S. Atherton, and J.E. Penner. "Atmospheric Optical Effects of Aerosols in Large Fires." In *Proceedings Ninth Conference on Fire and Forest Meteorology, April 21–24, 1987, San Diego, CA*, 253–57. Boston: American Meteorological Society, 1987.

Pruett, James. "Once in a Blue Moon." *Sky & Telescope*, March 1946.

Pyne, Stephen J. *Awful Splendor: A Fire History of Canada*. Vancouver: University of British Columbia Press, 2007.

Ramanathan, V.M., H. Agrawal, M. Akimoto, S. Aufhammer, L. Devotta, S.I. Emberson, M. Hasnain et al. *Atmospheric Brown Clouds: Regional Assessment Report with Focus on Asia*. Nairobi, KE: United Nations Environment Programme, 2008. http://www.unep.org/pdf/ABCSummaryFinal.pdf.

Reid, Gordon. *Around the Lower Peace*. Peace River, AB: Lower Peace Publishing Company, 1978.

Rittmaster, R., W.L. Adamowicz, B. Amiro, and R.T. Pelletier. "Economic Analysis of Health Effects from Forest Fires." *Canadian Journal of Forest Research* 36 (2006): 868–77.

———. "Erratum: Economic Analysis of Health Effects from Forest Fires." *Canadian Journal of Forest Research* 38 (2008): 908.

Salisbury, Harrison E. *The Great Black Dragon Fire: A Chinese Inferno*. Boston: Little Brown and Company, 1989.

Sandink, D. "The Resilience of the City of Kelowna: Exploring Mitigation before, during and after the Okanagan Mountain Park Fire." Research Paper Series no. 45, Institute for Catastrophic Loss Reduction, Toronto, 2009.

Sapkota, A., M. Symons, J. Kleissl, L. Wang, M.B. Parlange, J. Ondov, P.N. Breysse et al. "Impact of the 2002 Canadian Forest Fires on Particulate Matter Air Quality in Baltimore City." *Environmental Science and Technology* 39 (2005): 24–32.

Sausen, R., K. Gierens, M. Ponater, and U. Schumann. "A Diagnostic Study of the Global Distribution of Contrails. Part I: Present Day Climate." *Theoretical and Applied Climatology* 61 (1998): 127–41.

Schiermeier, Q. "Climate Change: That Sinking Feeling." *Nature* 435 (2005): 732–33. doi:10.1038/435732a.

Seitz, R. "Siberian Fire as 'Nuclear Winter' Guide." *Nature* 323 (September 11, 1986): 116–17.

Sinnott, Roger, Donald Olson, and Richard Fienberg. "What's a Blue Moon?" *Sky & Telescope*, May 1999.

Smith, C.D. "The Widespread Smoke Layer From Canadian Forest Fires During Late September 1950." *Monthly Weather Review* (September 1950): 180–84.

Smith, Jerry E. *Weather Warfare: The Military's Plan to Draft Mother Nature*. Kempton, IL: Adventures Unlimited Press, 2006.

Soja, Amber J., Brian Stocks, Paul Maczek, Mike Fromm, Rene Servranckx, Merritt Turetsky, and Brian Benscoter. "ARTAS: The Perfect Smoke." *Canadian Smoke Newsletter*, Fall 2008, 2–7. http://www.sk.lung.ca/canadian_smoke_newsletters/fall_2008.pdf.

Steward, F.R. "Heat Penetration in Soils beneath a Spreading Fire." Unpublished paper, Intermountain Forest and Range Experiment Station, USDA Forest Service, Missoula, MT, 1989.

Stocks, B.J., J.G. Goldammer, and L. Kondrashov. "Forest Fires and Fire Management in the Circumboreal Zone: Past Trends and Future Uncertainties." IMFN Discussion Paper no. 01, International Model Forest Network Secretariat, Natural Resources Canada, Ottawa, 2008.

Stocks, B.J., J.A. Mason, J.B. Todd, E.M. Bosch, B.M. Wotton, B.D. Amiro, M.D. Flannigan et al. "Large Forest Fires in Canada, 1959–1997." *Journal of Geophysical Research* 107, D1 (2002). doi:10.1029/2001JD000484.

Thistlethwaite, R.W. "Alberta–British Columbia Boundary: Positions and Descriptions of Monuments Established on Alberta–British Columbia Boundary." In *Monument Book*. Ottawa: Department of the Interior, 1951. Glenbow Archives, Calgary.

———. "Survey of Boundary Northward from the Termination of the 1923 Survey." In *Report and Field Journal, Summer Season 1950 and Winter Season 1950–1951*. Ottawa: Department of the Interior, 1952. Glenbow Archives, Calgary.

Todd, S.K., and H.A. Jewkes. "Wildland Fire in Alaska: A History of Organized Fire Suppression and Management in the Last Frontier." University of Alaska Fairbanks, Agricultural and Forestry Experiment Station, Bulletin no. 114, 2006.

Travis, D.J., A.M. Carleton, and R.G. Lauritsen. "Contrails Reduce Daily Temperature Range." *Nature* 418, no. 601 (August 8, 2002). doi:10.1038/418601a.

Turco, R.P., O.B. Toon, T.P. Ackerman, J.B. Pollack, and C. Sagan. "Nuclear Winter: Global Consequences of Multiple Nuclear Explosions." *Science* 222 (1983): 4630.

Turetsky, M.R., W. Donahue, and B.W. Benscoter. "Experimental Drying Intensifies Burning and Carbon Losses in a Northern Peatland." *Nature Communications* (November 2011). doi:10.1038/ncomms1523.

Udvardy, M.D.F. *Dynamic Zoogeography, with Special Reference to Land Animals*. New York: Van Nostrand Reinhold Co., 1969.

United Nations. "Fire Management—Global Assessment 2006: A Thematic Study Prepared in the Framework of the Global Forest Resources Assessment 2005." FAO Forestry Paper 151, United Nations, Rome, 2007. ftp://ftp.fao.org/docrep/fao/009/A0969E/A0969E00.pdf.

United Nations Environment Programme (UNEP), and Centre for Clouds, Chemistry, and Climate (C4). *The Asian Brown Cloud: Climate and Other Environmental Impacts. UNEP Assessment Report*. Pathumthani, TH: UNEP and C4, 2002. http://www.rrcap.unep.org/abc/impactstudy/.

United States Environmental Protection Agency. *Smog—Who Does It Hurt? What You Need to Know about Ozone and Your Health*. Washington, DC: United States Environmental Protection Agency, 1999. http://epa.gov/air/ozonepollution/pdfs/smog.pdf.

United States Geological Survey. "The Cataclysmic 1991 Eruption of Mount Pinatubo, Philippines." US Geological Survey Fact Sheet 113-97. Last modified February 28, 2005. http://pubs.usgs.gov/fs/1997/fs113-97/.

United States Government Accountability Office. "Technology Assessment: Protecting Structures and Improving Communications during Wildland Fires." Report to Congressional Requesters, GAO-05-380, 2005. http://www.gao.gov/assets/160/157598.html.

USDA Forest Service, Idaho Panhandle National Forests. *When the Mountains Roared: Stories of the 1910 Fire*. Coeur d'Alene, ID: USDA Forest Service, Idaho Panhandle National Forests, 1978.

Van de Hulst, Hendrik Christoffel. *Light Scattering by Small Particles*. New York: Dover Publications, 1981. Originally published by John Wiley & Sons, 1957.

Van Wagner, C.E. *Development and Structure of the Canadian Forest Fire Weather Index System*. Forestry Technical Report 35. Ottawa: Canadian Forestry Service, 1987.

Walters, Del, L.A. Snow, and Arnold Schwarzenegger. *2009 Wildfire Activity Statistics*. N.p.: California Department of Forestry and Fire Protection, 2009. http://www.fire.ca.gov/downloads/redbooks/2009/2009_Redbook_Complete_Final.pdf.

Ward, D.E. "Smoke from Wildland Fires." In *Health Guidelines for Vegetation Fire Events, Lima, Peru, October 6–9, 1998, Background Papers*, edited by Kee-Tai Goh, Dietrich Schwela, Johann G. Goldammer, and Orman Simpson, 70–85. N.p.: World Health Organization, 1999.

Watson, R.T., I.R. Noble, B. Bolin, N.H. Ravindranath, D.J. Verardo, and D.J. Dokken, eds. *Land Use, Land-Use Change, and Forestry. Special Report of the Intergovernmental Panel on Climate Change* (IPCC). Cambridge: Cambridge University Press, 2000.

Welke, S., and J. Fyles. "Organic Matter: Does It Matter?" SFMN Research Note Series no. 3, Sustainable Forest Management Network, University of Alberta, Edmonton, 2005.

Wells, C.G., R.E. Campbell, L.F. DeBano, C.E. Lewis, R.L. Fredriksen, E.C. Franklin, R.C. Froelich, and P.H. Dunn. "Effects of Fire on Soil, A State-of-Knowledge Review." General Technical Report WO-7, USDA Forest Service, Washington, DC. 1979.

Wexler, Harry. "The Great Smoke Pall of September 24–30, 1950." *Weatherwise*, December 1950, 129–34, 142.

Williams, J.H. "The Mega-Fire Phenomenon: Risk Implications for Land Managers and Policy-Makers." Spring Forest Industry Lecture Series, March 4, 2010, University of Alberta, Edmonton, http://www.ales.ualberta.ca/rr/SeminarsandLectures/ ForestIndustryLecture/Williams.aspx.

Wilson, C.C., and J.B. Davis. "Forest Fire Laboratory at Riverside and Fire Research in California: Past, Present, and Future." General Technical Report PSW-105, USDA Forest Service, Pacific Southwest Forest and Range Experiment Station, Berkeley, CA, 1988.

Wilson, Robert. "The Blue Sun of 1950 September." *Monthly Notices of the Royal Astronomical Society* 111 (1951): 478.

Winchester, Simon. *Krakatoa: The Day the World Exploded: August 27, 1883.* New York: HarperCollins, 2003.

Wright, J.G. "Forest-Fire Hazard Research: As Developed and Conducted at the Petawawa Forest Experiment Station." Forest Fire Hazard Paper no. 2, Division of Forest Protection, Forest Service, Department of the Interior, Ottawa, 1932.

Wright, J.G., and H.W. Beall. "The Application of Meteorology to Forest Fire Protection." Technical Communication no. 4. Imperial Forestry Bureau, Oxford, 1945.

Yalcin, K., C.P. Wake, K.J. Kreutz, M.S. Germani, and S.I. Whitlow. "Ice Core Paleovolcanic Records from the St. Elias Mountains, Yukon, Canada." *Journal of Geophysical Research* 112 (2007): D08102. doi:10.1029/2006JD007497.

INDEX

The Chinchaga Firestorm occurred from
September 20–22, 1950. These dates fall within
the range of the Chinchaga River Fire (June 1–
September 30, 1950). Because of this some
index entries are duplicated. Main entries
for dates are found at the top of the index.

16km firefighting limit
    attempts to remove, 11
    reporting fires within the, 130–31
    suppression policy within the, 9, 11–12, 15,
        19, 25, 132–34, 166
    timber within the, 11–12
1620–1920 dark days, 57–58, 59, 63–65
1780–1820 fire season, 58
1825 fire season
    fatalities, 58
    hectares burned, 58
1860s–1890s fire seasons
    fatalities, 62
    hectares burned, 62
1900–1919s fire seasons
    fatalities, 59, 61, 63

hectares burned, 59, 61, 63, 64
property damage or loss, 61, 63–64
western Siberia fires, 64–65
1920s–1930s fire seasons
    fatalities, 61
    firefighting costs, 62
    hectares burned, 4, 6
    lightning fires, 4, 6
    timber loss, 61–62
1940s fire season
    fatalities, 6–7, 129
    fire-related prosecutions, 120, 122
    game guardians converted to rangers, 132
    hectares burned, 7, 15, 62, 120, 121
    human-caused fires, 131–32
    let-burn policy, 131–33
    lightning fires, 14–15, 122
    number of fires, 7, 14, 121
    prescribed burns, 131–32
    primary causes of fires, 131–32
    timber loss, 120
1950s fire season. *See also* Chinchaga Firestorm;
        Chinchaga River Fire (1950)

Canadian Wildland Fire Strategy
   Declaration, 158
CANFIRE (boreal fire effects model), 102
carbon emissions, 92–93, 100–03, 164–65
carbon reserves in the boreal forest, xviii, xix
Carcajou Point, 36
Carlson, Henry, 44
Castle Rivery Valley fire (1934), 4, 6
Caterpillar Tractor Company, 144
cattle industry, 39
Ceanothus, xix–xx
Chandler, Craig, 93
Chapleau prescribed burn, 91
Chase, Charlie, 18
China, blue moon and blue sun
   sightings in, 82
Chinchaga Firestorm
   16km firefighting limit policy and the, 166
   air quality at ground level, 99
   animals displaced and killed, 38
   blame assigned for, 129
   Boundary Survey crew inside the, 117
   causal agents, other than wind, 59
   drought and the, 111–12
   extent of, 31, 51
   fatalities, 140
   flooding following, 38, 39
   growth simulation, 106, 113
   nitric acid formation, 87–88
   number of fires, xxiv, 10, 31, 33, 81, 101
   policy effects following, 166, 169
   rabies epidemic following, 38–39
   soil instability following, 38, 39
   spread, 107, 113
   water contamination following, 38
   wind and the, 26, 27–28, 51, 112–13, 117
Chinchaga Firestorm smoke. See also Black
   Sunday smoke pall
   Boundary Survey crew experiences, 101, 115
   distance travelled, 129
   duration of, 94
   health costs, 97
   in Keg River, 28, 101
   temperature effect, 54
   trajectories, 49, 86, 87–88
Chinchaga Lookout Tower, 15
Chinchaga River Fire (1950)

aerial reconnaissance on, 15–17
beginnings, 8–10
blame assigned, 9, 88, 129, 131
blocking pre-burn, 111–14
Boundary Survey crew inside the, 117
campfires as causal agents, 9, 25, 29
characteristics, 161
dates, 105
drought conditions in creating the, 107–09,
   111–12, 115, 117–18, 165–66
energy release rate, 145
events following reports of, 8
exclusion from the Forest Cover Map, 136
extent of, 8–9, 16–17, 25, 51, 105, 112–15
forest cover as causal agent, 109, 110
growth simulation modelling, 105–07
hectares burned, xxvii
intensity, 18
mapping the, 17, 114, 134
oil exploration, impact on, 18
personal accounts of, xxiv–xxvi
points of origin, 9–10
reports on the, 8, 128, 134
revegetation following, 18
soil instability following, 39
timber loss, 18
total area burned, 17
water contamination following, 18–19
Chinchaga River Fire (1950), spread of
dates, 113
major fire activity, 106, 107
major runs, 113
path, 18
pilot reports of, 16
projected, 114
rates of, 111
smoke arrives at Keg River, 17
spruce cones rain fire from the sky,
   19, 21–22, 34
suppression/control policy, 9, 18
wind and the, 16, 19, 21–22, 112–15
Chinchaga River Fire (2001), 166
Chinchaga River flooding, 39
Chinchaga River Valley, 4, 11
Chisholm Fire, 85
Christian, L., 28
cigarettes/cigars thrown from planes, 123–24

circumboreal forests, largest
    wildfires in, 65–66
Clearwater Forest Reserve, 3, 6
Cleveland, Ohio, 129
climate change
    carbon emissions and, 92–93, 101
    "Climate Changes Endanger World's Food
        Output" (Schmeck), 90
    "The Cooling World" (Gwynne), 90
    Intergovernmental Panel on Climate
        Change Fourth Assessment Report, 164
    Krakatoa volcano eruption, 67
    managing, 67, 165–66
    mega fires and, 164
    mosquito-borne illnesses from, 98
    nuclear winter theory, 91
    preparedness, 146–47
    processes contributing to, 165
"Climate Changes Endanger World's Food
    Output" (Schmeck), 90
Cloud-Aerosol Lidar and Infrared Pathfinder
    Satellite Observation (CALIPSO), 92, 102
cloud seeding, 45–46, 89
coagulation effect on blue moon and blue sun
    sightings, 79–80, 81
color, factors influencing, 76–78
community protection zones, 99, 131
conservation
    Eastern Rockies Forest Conservation Area
        (ERFCA), 3, 6, 14–15, 122, 167–68
    Eastern Rockies Forest Conservation Board
        (ERFCB), 3, 6, 12, 14
    Eastern Rocky Mountain Forest
        Conservation Act, 6, 12, 14
contrails, 89
convection driven fire, 148
"The Cooling World" (Gwynne), 90
Coos Fire (1868), 62
cougars, bounties on, 125
Coulson Aircrane Ltd., 152
counter fire, 149
coyotes, deaths from poisoning, 39
Cree, ecological use of fire, xxi–xxii
crown fires, xxi, 66, 159
Crowsnest Forest Reserve, 3, 6
Crutzen, Paul, 93
cultural use of fire, xviii, xxi–xxiii

Cypress Hills Forest Reserve, 6, 148
dark days. See also Black Sunday smoke pall
    1620–1920, 57–58, 59, 63–65
    Krakatoa volcano eruption, 66
    Lac La Biche Fire, 63–64
    noteworthy events associated with, 57
    speculation on causes of, 58
    when the moon and the sun turned blue,
        xxiii, xxv–xxvi, 51, 67, 69–72, 75–77, 79–82
Dead Out, 122
Delahey, Wallace, 11
Derocher, Wilfred, 7
desert dust, blue and green suns from, 82
Desjarlais, Theresa, 64
detection, protection program
    recommendation, 12
dinosaur bone excavation, 125
Dinosaur Discovery Center, 125
Doig Lookout Tower, 15
Dominion Forest Service, 6, 131, 166
Drayton Valley fires, 32
drought
    Chinchaga Firestorm and the, 111–12
    Chinchaga River Fire and the, 107–09,
        111–12, 115, 117–18, 165–66
    El Niño and, 165
    Keg River area, 111
    Richardson Fire and the, 166
dry cold front, 115

earth-orbiting needles, 78
Eastern Rockies Forest Conservation Area
    (ERFCA), 3, 6, 14–15, 122, 167–68
Eastern Rockies Forest Conservation Board
    (ERFCB), 3, 6, 12, 14
Eastern Rocky Mountain Forest
    Conservation Act, 6
Eastern Slopes Forest Reserves, 6, 7
ecological role of wildfires, xviii–xxii, 124,
    160, 161, 162
economics. See also financial impacts of fire
    costs of air pollution, 95, 96, 97–98
    of extinguishment, 15
    of fighting fire, 62, 146–47
    of suppression, 130, 146, 164
Edgecombe, Harry, 18
Edmonton, 4, 25

Hack, Dan, 113
Haileybury Fire (1922), 61
Hammerstedt, Ron, xxvii, 114
Hanson, K.H. "Slim," 19
Harpe, William, 6–7
Harrington, D.B., 129
Harris Fire, 152
Harvie, J., 131
Hay Lake, 4, 16
Hay River, 11, 117
Hay River Valley, 117
hazard management, 167–68
hazard management research, 166–67
health. *See also* fatalities
    air pollution's effect on, 83–84, 94–103
    settlers, fire-related, 22, 28
helicopters, 142
Henry, Robert, 64
Highwood Fire (1936), 6
Hinckley Fire (1894), 62, 63
Hiscock, Philip, 73
Holdsworth, Gerald, 86–87
Holman, Harry, 124
Holt, Benjamin, 143–44
homeowners
    fire prevention responsibility, 157–58
    interior smoke, 25, 100k
    stay and defend vs. leave early
        strategy, 32, 34
Horn Plateau Fire (1995), 160
Hornsby, Richard, 143–44
Hourglass Fire (2006), 124–25
House River Fire (2002), 160
Hudson's Bay Company, 36, 40, 168
Huestis, Donna, 20
Huestis, Eric, xxiv, xxv–xxvi, 9, 11, 19–21,
    119–22, 127–29, 131–33, 139, 155, 169
Huestis, Ivy, 19–20
Huffman, Donald, 81
Hughes, Howard, 150
human-caused fires
    agricultural use of, xviii, xxiii, 121, 122–
        23, 125, 129
    campfires, 9, 29, 122, 132
    eliminating. *see* prevention
    greater than 200 hectares, 33
    increases in, 130

indigenous peoples of the boreal forest,
        xviii, xxi–xxiii
    lit cigarettes/cigars thrown from
        planes, 123–24
    percent of total, 123
    present-day landscape from, 136
    prosecution for, 120, 122–23, 156
    prospectors, 124
    settlers, 10, 119, 120, 122–23
    suppression policy, 166
    trappers, 124
hunter-gatherer lifestyle, xxii

Imperial Oil "Last Chance" well, 135
India, 1998 biomass burning fatalities, 98
indigenous peoples of the boreal forest
    cultural and natural fire regimes,
        xviii, xxi–xxiii
    ecological use of fire, xxi–xxii, 124, 161
    fire controls, xxii
    medicinal use of fire, 124
    wildfires caused by, 124
Indonesia, 1997 fires, 100, 165
Institute of Forestry Standing Committee on
        Forest Fire, 144
interface fires, 125, 127, 144–46, 154–55
Intergovernmental Panel on Climate Change
        Fourth Assessment Report, 164
Iosegun Fire, 30
Isle of Man, 51

jack pine, xxi
Jackson, Frank, xxiv–xxvi, 17–18, 23, 28–29, 32,
        34–38, 169
Jackson, Louis, 19, 21, 32
Jackson, Louise, 36, 37
Jackson, Mary (neé Percy), xxiv–xxv, 17, 22–23,
        25, 34–38, 71–72
Jackson barley, 37
Jackson homestead, 22–23, 25
Jansen, J.L., 120
Japanese bombing strategy, 90
Jaworski, Jack, 30–31, 55
Johnson, B.H., 114
Jones, West, 70
Joy, George C., 61–62, 66

Other Titles from
The University of Alberta Press

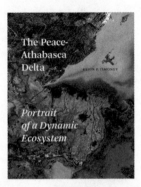

THE PEACE-ATHABASCA DELTA
Portrait of a Dynamic Ecosystem
Kevin P. Timoney
608 pages • Wall map plus over 450 figures: maps,
illustrations, graphs, charts, photographs, aerial
photographs, notes, bibliography, appendices, index; colour
throughout
978-0-88864-603-3 | $160.00 (T) cloth
978-0-88864-730-6 | $90.00 (T) paper
978-0-88864-802-0 | $72.00 (T) PDF
Nature/Ecology | Rivers | Petroleum Industry

THE ALGAL BOWL
Overfertilization of the World's Freshwaters and Estuaries
David W. Schindler & John R. Vallentyne
348 pages | B&W photographs, colour section, illustrations,
maps, tables, graphs, glossary, index
978-0-88864-484-8 | $34.95 (T) paper
Water Management | Ecology
Copublished with Earthscan

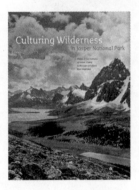

CULTURING WILDERNESS IN JASPER NATIONAL PARK
Studies in Two Centuries of Human History in the Upper Athabasca
River Watershed
I. S. MacLaren, Michael Payne, Peter J. Murphy, PearlAnn
Reichwein, Lisa McDermott, C. J. Taylor, Gabrielle Zezulka-
Mailloux, Zac Robinson and Eric Higgs
The Rt. Hon. Jean Chrétien, Foreword
400 pages| Colour throughout, introduction, foreword, maps,
notes, bibliography, index
Mountain Cairns: A series on the history and culture of the
Canadian Rocky Mountains
978-0-88864-483-1 | $45.00 (T) paper
978-0-88864-570-8 | $35.99 (T) PDF
History | Tourism | National Parks